ENCYCLOPAEDIA OF ENVIRONMENTAL BIOTECHNOLOGY

ENCYCLOPAEDIA OF ENVIRONMENTAL BIOTECHNOLOGY

Vol. 1

Shahzad Ahmad

ANMOL PUBLICATIONS PVT. LTD.
NEW DELHI - 110 002 (INDIA)

ANMOL PUBLICATIONS PVT. LTD.
H.O.: 4374/4B, Ansari Road, Darya Ganj,
New Delhi-110 002 (India)
Ph.: 23278000, 23261597
B.O.: No. 1015, Ist Main Road, BSK IIIrd Stage
IIIrd Phase, IIIrd Block,
Bangalore - 560 085 (India)
Visit us at: www.anmolpublications.com

Encyclopaedia of Environmental Biotechnology

First Published, 2008

ISBN 978-81-261-3539-4 (Set)

PRINTED IN INDIA

Printed at Mehra Offset Press, Delhi.

Contents

Preface

Environmental biotechnology is the applied technology for the study of natural environment. Environmental biotechnology could also imply that one tries to harness biological process for commercial use and exploitation. Environmental biotechnology, by contrast, deals with farless readily apparently dramatic topics, and though their importance, albeit different, may be every bit as great their direct relevance is farless readily appreciated by the bulk of population. Cleaning up contamination and dealing rationally with waste is of course, in everybody's best interests, but for most people, this is like simply addressing a problem. Some of the potentially beneficial uses of biological engineering, which touch the lives of the majority of people, however directly involve much simpler approaches. In many respects, there is a logical link between the aims of environmental biotechnology.

Environmental biotechnology is fundamentally rooted in wastes, in its various guises, typically being concerned with the remediation of the contamination caused by previous use and the impact reduction of current activity or the control of pollution. Thus, the principal aim of this stream are the manufacture of products in environmentally harmonious manner, which allow for the minimisation of harmful solids, liquids or gaseous outputs or clean up the residual effects of earlier human occupations. Environmental biotechnologists may enhance or optimise conditions for existing biological systems to make their activities happen faster or more efficiently or they resort to some form of alteration to bring about the desired results. The variety of organism, which may play a part in environmental applications of biotechnology is wide and ranging from microbes through trees and all are utilised on one of the same three fundamental bases.

Accordingly, the range of businesses to which environmental biotechnology has potential relevance is almost limitless. All commercial operations generate wastes in one form or the other. Additionally, biotechnology can be of further commercial significance by converting low-cost organic feed stocks into high-value products.

Biotechnology based methods are seen as essential to help meet European Union (EU) targets for biowaste diversion and reductions in pollutants. Across the world, the existing regulations on environmental pollution are required to be rigorously enforced, with more stringent compliance standards implemented. All of this is expected to stimulate the sale of biotechnology based environmental processing methods significantly and in particular, the global market share is projected to grow faster than the general biotic sector trends.

This exhaustive work has a perfect treatment with environmental biotechnology in an objective manner. Expectedly, this work would be capable to create a sort of awareness among the people and make the study of environmental biotechnology easier and more meaningful.

Editor

1

Introduction

Environmental biotechnology is applied and used to study the natural environment. Environmental biotechnology could also imply that one try to harness biological process for commercial uses and exploitation. The International Society of Environmental Biotechnology defines environmental biotechnology as "the development, use and regulation of biological systems for remediation of contaminated environments (land, air, water), and for environment-friendly processes (green manufacturing technologies and sustainable development)".

"Environmental Biotechnology can be significant towards Industrial Biotechnology."

The *"Chambers Science and Technology Dictionary"* defines biotechnology as the use of organisms or their components in industrial or commercial processes, which can be aided by the techniques of genetic manipulation in developing, e.g. novel plants for agriculture or industry. Despite the inclusiveness of this definition, the biotechnology sector is still often seen as largely medical or pharmaceutical in nature, particularly amongst the general public. While to some extent the huge research budgets of the drug companies and the widespread familiarity of their products makes this understandable, it does distort the full picture and somewhat unfairly so. However, while therapeutic instruments form, in many respects, the 'acceptable' face of biotechnology, elsewhere the science is all too frequently linked with unnatural interference. While the agricultural, industrial and environmental applications of biotechnology are potentially very great, the shadow of Frankenstein has often been cast across them. Genetic engineering may be relatively commonplace in pharmaceutical thinking and yet in other spheres, like agriculture for example, society can so readily and thoroughly demonise it.

The history of human achievement has always been episodic. For a while, one particular field of endeavour seems to hold sway as the preserve of genius and development, before the focus shifts and development forges ahead in dizzy exponential rush in an entirely new direction. So it was with art in the renaissance, music in the 18th century, engineering in the 19th and physics in the 20th. Now it is the age of the biological, possibly best viewed almost as a rebirth, after the great heyday of the Victorian naturalists, who provided so much input into the developing science.

It is then, perhaps, no surprise that the European Federation of Biotechnology begins its 'Brief History' of the science in the year 1859, with the publication of *"On the Origin of Species by Means of Natural Selection"* by Charles Darwin. Though his famous voyage aboard *HMS Beagle*, which led directly to the formulation of his (then) revolutionary ideas, took place when he was a young man, he had delayed making them known until 1858, when he made a joint presentation before the Linnaean Society with Alfred Russell Wallace, who had, himself, independently come to very similar conclusions. Their contribution was to view evolution as the driving force of life, with successive selective pressures over time endowing living beings with optimised characteristics for survival. Neo-Darwinian thought sees the interplay of mutation and natural selection as fundamental. The irony is that Darwin himself rejected mutation as too deleterious to be of value, seeing such organisms, in the language of the times, as 'sports' – oddities of no species benefit. Indeed, there is considerable evidence to suggest that he seems to have espoused a more Lamarckist view of biological progression, in which physical changes in an organism's lifetime were thought to shape future generations. Darwin died in 1882. Ninety-nine years after his death, the first patent for a genetically modified organism was granted to Ananda Chakrabarty of the US General Electric, relating to a strain of *Pseudomonas aeruginosa* engineered to express the genes for certain enzymes in order to metabolise crude oil.

Twenty years later still, in the year that saw the first working draft of the human genome sequence published and the announcement of the full genetic blueprint of the fruit fly, *Drosophila melanogaster*, that archetype of eukaryotic genetics research, biotechnology has become a major growth industry with increasing numbers of companies listed on the world's stock exchanges. Thus, at the other end of the biotech timeline, a century and a half on from *Origin of Species*, the principles it first set out remain of direct relevance for what has been termed the 'chemical evolution' of biologically active substances and are commonly used in laboratories for *in vitro* production of desired qualities in biomolecules.

Role of Environmental Biotechnology

While pharmaceutical biotechnology represents the glamorous end of the market, environmental applications are decidedly more in the Cinderella mould. The reasons for this are fairly obvious. The prospect of a cure for the many diseases and conditions currently promised by gene therapy and other biotech-oriented medical miracles can potentially touch us all. Our lives may, quite literally, be changed. Environmental biotechnology, by contrast, deals with far less apparently dramatic topics and, though their importance, albeit different, may be every bit as great, their direct relevance is far less readily appreciated by the bulk of the population. Cleaning up contamination and dealing rationally with wastes is, of course, in everybody's best interests, but for most people, this is simply addressing a problem which they would rather had not existed in the first place.

Even for industry, though the benefits may be noticeable on the balance sheet, the likes of effluent treatment or pollution control are more of an inevitable obligation than a primary goal in themselves. In general, such activities are typically funded on a distinctly limited budget and have traditionally been viewed as a necessary inconvenience. This is in no way intended to be disparaging to industry; it simply represents commercial reality.

In many respects, there is a logical fit between this thinking and the aims of environmental biotechnology. For all the media circus surrounding the grand questions of our age, it is easy to forget that not all forms of biotechnology involve xenotransplantation, genetic modification, the use of stem cells or cloning.

Some of the potentially most beneficial uses of biological engineering, and which may touch the lives of the majority of people, however indirectly, involve much simpler approaches. Less radical and showy, certainly, but powerful tools, just the same. Environmental biotechnology is fundamentally rooted in waste, in its various guises, typically being concerned with the remediation of contamination caused by previous use, the impact reduction of current activity or the control of pollution. Thus, the principal aims of this field are the manufacture of products in environmentally harmonious ways, which allow for the minimisation of harmful solids, liquids or gaseous outputs or the clean-up of the residual effects of earlier human occupation.

The means by which this may be achieved are essentially twofold. Environmental biotechnologists may enhance or optimise conditions for existing biological systems to make their activities happen faster or more efficiently, or they resort to some form of alteration to bring about the desired outcome. The variety of organisms which may play a part in environmental applications of biotechnology is huge, ranging from microbes through to trees and all are utilised on one of the same three fundamental bases – accept, acclimatise or alter. For the vast majority of cases, it is the former approach, accepting and making use of existing species in their natural, unmodified form, which redominates.

Scope in Practice

There are three key points for environmental biotechnology interventions, namely in the manufacturing process, waste management or pollution control. Accordingly, the range of businesses to which environmental biotechnology has potential relevance is almost limitless. One area where this is most apparent is with regard to waste. All commercial operations generate waste of one form or another and for many, a proportion of what is produced is biodegradable. With disposal costs rising steadily across the world, dealing with refuse constitutes an increasingly high contribution to overheads. Thus, there is a clear incentive for all businesses to identify potentially cost-cutting approaches to waste.

The three intervention points employ them where possible. Changes in legislation throughout Europe, the US and elsewhere, have combined to drive these issues higher up the political agenda and biological methods of waste treatment have gained far greater acceptance as a result. For those industries with particularly high biowaste production, the various available treatment biotechnologies can offer considerable savings.

Manufacturing industries can benefit from the applications of whole organisms or isolated biocomponents. Compared with conventional chemical processes, microbes and enzymes typically function at lower temperatures and pressures. The lower energy demands this makes leads to reduced costs, but also has clear benefits in terms of both the environment and workplace safety. Additionally, biotechnology can be of further commercial significance by converting low-cost organic feedstocks into high value products or, sinc enzymatic reactions are more highly specific than their chemical counterparts, by deriving final substances of high relative purity. Almost inevitably, manufacturing companies produce wastewaters or effluents, many of which contain biodegradable contaminants, in varying degrees.

Though traditional permitted discharges to sewer or watercourses may be adequate for some, other industries, particularly those with recalcitrant or highly concentrated effluents, have found significant benefits to be gained from using biological treatment methods themselves on site. Though careful monitoring and process control are essential, biotechnology stands as a particularly cost-effective means of reducing the pollution potential of wastewater, leading to enhanced public relations, compliance with environmental legislation and quantifiable cost-savings to the business.

Those involved in processing organic matter, for example, or with drying, printing, painting or coating processes, may give rise to the release of volatile organic compounds (VOCs) or odours, both of which represent environmental nuisances, though the former is more damaging than the latter. For many, it is not possible to avoid producing these emissions altogether, which leaves treating them to remove the offending contaminants the only practical solution. Especially for relatively low concentrations of readily water-soluble VOCs or odorous chemicals, biological technologies can offer an economic and effective alternative to conventional methods.

The use of biological cleaning agents is another area of potential benefit, especially where there is a need to remove oils and fats from process equipment, work surfaces or drains. Aside from typically reducing energy costs, this may also obviate the need for toxic or dangerous chemical agents. The pharmaceutical and brewing industries, for example, both have a long history of employing enzyme-based cleaners to remove organic residues from their process equipment. In addition, the development of effective biosensors, powerful tools which rely on biochemical reactions to detect specific substances, has brought benefits to a wide range of sectors, including the manufacturing, engineering, chemical, water, food and beverage industries. With their ability to detect even small amounts of their particular target chemicals, quickly, easily and accurately, they have been enthusiastically adopted for a variety of process monitoring applications, particularly in respect of pollution assessment and control.

Contaminated land is a growing concern for the construction industry, as it seeks to balance the need for more houses and offices with wider social and environmental goals. The reuse of former industrial sites, many of which occupy prime locations, may typically have associated planning conditions attached which demand that the land be cleaned up as part of the development process. With urban regeneration and the reclamation of 'brown-field' sites increasingly favoured in many countries over the use of virgin land, remediation has come to play a significant role and the industry has an ongoing interest in identifying cost-effective methods of achieving it. Historically, much of this has involved simply digging up the contaminated soil and removing it to landfill elsewhere. Bioremediation technologies provide a competitive and sustainable alternative and in many cases, the lower disturbance allows the overall scheme to make faster progress.

As the previous brief examples show, the range of those which may benefit from the application of biotechnology is lengthy and includes the chemical, pharmaceutical, water, waste management and leisure industries, as well as manufacturing, the military, energy generation, agriculture and horticulture. Clearly, then, this may have relevance to the viability of these ventures and, as was mentioned at the outset, biotechnology is an essentially commercial activity. Environmental biotechnology must compete in a world governed by the *best practicable environmental option* (BPEO) and the *best available techniques not entailing excessive cost* (BATNEEC). Consequently, the economic aspect will always have a large influence on the uptake of all initiatives in environmental biotechnology and, most particularly, in the selection of methods to be used in any given situation. It is impossible to divorce this context from the decision making process. By the same token, the sector itself has its own implications for the wider economy.

Market for Environmental Biotechnology

The UK's Department of Trade and Industry estimated that 15-20 per cent of the global environmental market in 2001 was biotech-based, which amounted to about 250-300 billion US dollars and the industry is projected to grow by as much as tenfold over the following five years. This expected growth is due to greater acceptance of biotechnology for clean manufacturing applications and energy production, together with increased landfill charges and legislative changes in waste management which also alter the UK financial base favourably with respect to bioremediation. Biotechnology-based methods are seen as essential to help meet European Union (EU) targets for biowaste diversion from landfill and reductions in pollutants. Across the world the existing regulations on environmental pollution are predicted to be more rigorously enforced, with more stringent compliance standards implemented. All of this is expected to stimulate the sales of biotechnology-based environmental processing methods significantly and, in particular, the global market share is projected to grow faster than the general biotech sector trend, in part due to the anticipated large-scale EU aid for environmental clean-up in the new accession countries of Eastern Europe.

Other Sources

The Bioindustry Association (BIA) survey, *Industrial Markets for UK Biotechnology – Trends and Issues,* published in 1999 does not quote any monetary sector values per year, but gives the size of the UK sector as employing 40,000 people in 1998 with an average yearly growth over 1995-98 of 20 per cent. Environmental biotech is reported as representing around 10 per cent of this sector. An Arthur Anderson report of 1997 gives the turnover of the UK biotech sector as 702 million pounds sterling in 1995/96, with a 50 per cent growth over three years. A 1998 Ernst and Young report on the European Life.

Sciences Sector says that the market for biotechnology products has the potential to reach 100 billion pounds sterling worldwide by 2005. The Organisation for Economic Cooperation and Development (OECD) estimates that the global market for environmental biotechnology products and services alone will rise to some US$75 billion by the year 2000, accounting for some 15 to 25 per cent of the overall environmental technology market, which has a growth rate estimated at 5.5 per cent per annum. The UK potential market for environmental biotechnology products and services is estimated at between 1.65 and 2.75 billion US dollars and the growth of the sector stands at 25 per cent per annum, according to the *BioCommerce Data European Biotechnology Handbook*. An unsourced quote found on a Korean The world market size of biotechnology products and services was estimated to be approximately 390 billion US dollars in the year 2000.

The benefits are not, however, confined to the balance sheet. The Organisation for Economic Cooperation and Development (OECD 2001) concluded that the industrial use of biotechnology commonly leads to increasingly environmentally harmonious processes and additionally results in lowered operating and/or capital costs. For years, industry has appeared locked into a seemingly unbreakable cycle of growth achieved at the cost of environmental damage. The OECD investigation provides what is probably the first hard evidence to support the reality of biotechnology's long-heralded promise of alternative production methods, which are ecologically sound and economically efficient. A variety of industrial sectors including pharmaceuticals, chemicals, textiles, food and energy were examined, with a particular emphasis on biomass renewable resources, enzymes and biocatalysis. While such approaches may have to be used in tandem with other processes for maximum effectiveness, it seems that their use invariably leads to reduction in operating or capital costs, or both. Moreover, the research also concludes that it is clearly in the interests of governments of the developed and developing worlds alike to promote the use of biotechnology for the substantial reductions in resource and energy consumption, emissions, pollution and waste production it offers. The potential contribution to be made by the appropriate use of biotechnology to environmental and economic sustainability would seem to be clear.

The upshot of this is that few biotech companies in the environmental sector perceive problems for their own business development models, principally as a result of the wide range of businesses for which their services are applicable, the relatively low market penetration to date and the large potential for growth. Competition within the sector is not seen as a major issue either, since the field is still largely open and unsaturated. Moreover, there has been a discernible tendency in recent years towards niche specificity, with companies operating in more specialised subarenas within the environmental biotechnology umbrella. Given the number and diversity of such possible slots, coupled with the fact that new opportunities, and the technologies to capitalise on them, are developing apace, this trend seems likely to continue. It is not without some irony that companies basing their commercial activities on biological organisms should themselves come to behave in such a Darwinian fashion. However, the picture is not entirely rosy.

Typically the sector comprises a number of relatively small, specialist companies and the market is, as a consequence, inevitably fragmented. Often the complexities of individual projects make the application of 'standard' off the shelf-approaches very difficult, the upshot being that much of what is done must be significantly customised. While this, of course, is a strength and of great potential environmental benefit, it also has hard commercial implications which must be taken into account. A sizeable proportion of companies active in this sphere, have no products or services which might reasonably be termed suitable for generalised use, though they may have enough expertise, experience or sufficiently perfected techniques to deal with a large number of possible scenarios.

The fact remains that one of the major barriers to the wider uptake of biological approaches is the high perceived cost of these applications. Part of the reason for this lies in historical experience. For many years, the solutions to all environmental problems were seen as expensive and for many, particularly those unfamiliar with the multiplicity of varied technologies available, this has remained the prevalent view. Generally, there is often a lack of financial resource allocation available for this kind of work and biotech providers have sometimes come under pressure to reduce the prices for their services as a result. Greater awareness of the benefits of biotechnology, both as a means to boost existing markets and for the opening up of new ones, is an important area to be addressed.

Many providers, particularly in the UK, have cited a lack of marketing expertise as one of the principal barriers to their exploitation of novel opportunities. In addition, a lack of technical understanding of biotech approaches amongst target industries and, in some cases, downright scepticism regarding their efficacy, can also prove problematic. Good education, in the widest sense, of customers and potential users of biological solutions will be one major factor in any future upswing in the acceptance and utilisation of these technologies.

Modalities and Influences

Another of the key factors affecting the practical uptake of environmental biotechnology is the effect of local circumstances. Contextual sensitivity is almost certainly the single most important factor in technology selection and represents a major influence on the likely penetration of biotech processes into the marketplace. Neither the nature of the biological system, nor of the application method itself, play anything like so relevant a role. This may seem somewhat unexpected at first sight, but the reasons for it are obvious on further inspection.

While the character of both the specific organisms and the engineering remain essentially the same irrespective of location, external modalities of economics, legislation and custom vary on exactly this basis. Accordingly, what may make abundant sense as a biotech intervention in one region or country, may be totally unsuited to use in another. In as much as it is impossible to discount the wider global economic aspects in the discussion, disassociating political, fiscal and social conditions equally cannot be done, as the following example illustrates. In 1994, the expense of bioremediating contaminated soil in the United Kingdom greatly exceeded the cost of removing it to landfill. Six years later, with successive changes of legislation and the imposition of a landfill tax, the situation has almost completely reversed. In those other countries where landfill has always been an expensive option, remediation has been embraced far more readily.

While environmental biotechnology must, inevitably, be viewed as contextually dependent, as the previous example shows, contexts can change. In the final analysis, it is often fiscal instruments, rather than the technologies, which provide the driving force and sometimes seemingly minor modifications in apparently unrelated sectors can have major ramifications for the application of biotechnology.

Again as has been discussed, the legal framework is another aspect of undeniable importance in this respect. Increasingly tough environmental law makes a significant contribution to the sector and changes in regulatory legislation are often enormously influential in boosting existing markets or creating new ones. When legislation and economic pressure combine, as, for example, they have begun to do in the European Landfill Directive, the impetus towards a fundamental paradigm shift becomes overwhelming and the implications for relevant biological applications can be immense.

There is a natural tendency to delineate, seeking to characterise technologies into particular categories or divisions. However, the essence of environmental biotechnology is such that there are many more similarities than differences. Though it is, of course, often helpful to view individual technology uses as distinct, particularly when considering treatment options for a given environmental problem, there are inevitably recurrent themes which feature throughout the whole topic. Moreover, this is a truly applied science. While the importance of the laboratory bench cannot be denied, the controlled world of research translates imperfectly into the harsh realities of commercial implementation.

Thus, there can often be a dichotomy between theory and application and it is precisely this fertile ground which is explored in the present work. In addition, the principal underlying approach of specifically *environmental* biotechnology, as distinct from other kinds, is the reliance on existing natural cycles, often directly and in an entirely unmodified form. Thus, this science stands on a foundation of fundamental biology and biochemistry. To understand the application, the biotechnologist must simply examine the essential elements of life, living systems and ecological circulation sequences. However, engineered the approach, this fact remains true.

In many respects, environmental biotechnology stands as the purest example of the newly emergent bioindustry, since it is the least refined, at least in terms of the basis of its action. In essence, all of its applications simply encourage the natural propensity of the organisms involved, while seeking to enhance or accelerate their action. Hence, optimisation, rather than modification, is the typical route by which the particular desired end result, whatever it may be, is achieved and, consequently, a number of issues feature as common threads within the discussions of individual technologies.

Integrated Approach

Integration is an important aspect for environmental biotechnology. One theme that will be developed throughout this book is the potential for different biological approaches to be combined within treatment trains, thereby producing an overall effect which would be impossible for any single technology alone to achieve. However, the wider goal of integration is not, of necessity, confined solely to the specific methods used. It applies equally to the underpinning knowledge that enables them to function in the first place and an understanding of this is central to the rationale behind this book. In some spheres, traditional biology has become rather unfashionable and the emphasis has shifted to more exciting sounding aspects of life science.

While the new-found concentration on 'ecological processes', or whatever, sounds distinctly more 'environmental', in many ways, and somewhat paradoxically, it sometimes serves the needs of environmental biotechnology rather less well. The fundamentals of living systems are the stuff of this branch of science and, complex though the whole picture may be; at its simplest the environmental biotechnologist is principally concerned with a relatively small number of basic cycles. In this respect, a good working knowledge of biological processes like respiration, fermentation and photosynthesis, a grasp of the major cycles by which carbon, nitrogen and water are recycled and an appreciation of the flow of energy through the biosphere must be viewed as prerequisites.

Unsurprisingly, then, these basic processes appear throughout this book, either explicitly or tacitly accepted as underpinning the context of the discussion. The intent here has been neither to insult the readership by parading what is already well known, nor gloss over aspects which, if left unexplained, at least in reasonable detail, might only serve to confuse. However, this is expressly not designed to be a substitute for much more specific texts on these subjects, nor an entire alternative to a cohesive course on biology or biochemistry. The intention is to introduce and explain the necessary aspects and elements of various metabolic pathways, reactions and abilities as required advancing the reader's understanding of this particular branch of biotechnology.

A large part of the reasons for approaching the subject in this way is the fact that there really is no such thing as a 'typical' environmental biotechnologist. Practitioners come into the profession from a wide variety of disciplines and by many different routes. Thus, amongst their ranks are agronomists, biochemists, biologists, botanists, enzymologists, geneticists, microbiologists, molecular biologists, process engineers and protein technologists, all of whom bring their own particular skills, knowledge base and experiences.

The applied nature of environmental biotechnology is obvious. While the science underlying the processes themselves may be as pure as any other, what distinguishes this branch of biological technology are the distinctly real-life purposes to which it is put. Hence, part of the intended function of this book is to attempt to elucidate the former in order to establish the basis of the latter.

At the same time, as any applied scientist will confirm, what happens in the field under operational conditions represents a distinct compromise between the theoretical and the practically achievable. At times, anything more than an approximation to the expected results may be counted as something of a triumph of environmental engineering.

The celebrated astronomer and biologist, Sir Fred Hoyle, said that the solutions to major unresolved problems should be sought by the exploration of radical hypotheses, while simultaneously adhering to well-tried and tested scientific tools and methods. This approach is particularly valid for environmental biotechnology. With new developments in treatment technologies appearing all the time, the list of what can be processed or remediated by biological means is ever changing. By the same token, the applications for which biotechnological solutions are sought are also subject to alteration.

For the biotech sector to keep abreast of these new demands it may be necessary to examine some truly 'radical hypotheses' and possibly make use of organisms or their derivatives in ways previously unimagined.

This is the basis of innovation; the inventiveness of an industry is often a good measure of its adaptability and commercial robustness.

2

Significant Aspects of Biotechnology

Biotechnology is technology based on biology, especially when used in agriculture, food science, and medicine. The United Nations Convention on Biological Diversity has come up with one of many definitions of biotechnology:

> "Biotechnology means any technological application that uses biological systems, living organisms, or derivatives thereof, to make or modify products or processes for specific use."

Before the 1970s, the term, *biotechnology*, was primarily used in the food processing and agriculture industries. Since the 1970s, it began to be used by the Western scientific establishment to refer to laboratory-based techniques being developed in biological research, such as recombinant DNA or tissue culture-based processes. In fact, the term should be used in a much broader sense to describe the whole range of methods, both ancient and modern, used to manipulate organic matter to meet human needs. So the term can be defined as, "The application of indigenous and/or scientific knowledge to the management of (parts of) microorganisms, or of cells and tissues of higher organisms, so that these supply goods and services of use to human beings.

There has been a great deal of talk - and money - poured into biotechnology with the hope that miracle drugs will appear. While there do seem to be a small number of efficacious drugs, in general, the biotech revolution has not happened in the pharmaceutical sector. However, recent progress with monoclonal antibody based drugs, such as Genentech's Avastin (tm) suggest that biotech may finally have found a role in pharmaceutical sales.

Biotechnology combines disciplines like genetics, molecular biology, biochemistry, embryology and cell biology, which are in turn linked to practical disciplines like chemical engineering, information technology, and robotics.

History of Biotechnology

Biotechnology is the application of scientific and engineering principles to the processing of materials by biological agents to provide goods and services. From its inception, biotechnology has maintained a close relationship with society. Although now most often associated with the development of remarkable drugs, historically biotechnology has been principally associated with food, addressing such issues as malnutrition and famine. The history of biotechnology begins with zymotechnology, which commenced with a focus on brewing techniques for beer. By World War I, however, zymotechnology would expand to tackle larger industrial issues, and the potential of industrial fermentation gave rise to biotechnology.

The belief that the needs of a society could be met by fermenting industrial waste was an important ingredient of the "chemurgic movement," which began in the 1930s. Expectations were raised during the 1960s by a process that grew single-cell protein from petroleum oil. This new product was seen to be a solution to growing food shortages and world hunger. In the 1970s, biotechnology offered a possible solution to the escalation in oil prices and the increasing energy demands by synthesising a new product called gasohol. However, both the single-cell protein and gasohol projects failed to progress due to varying issues including public resistance, a changing economic scene, and shifts in political power.

Yet the formation of a new field, genetic engineering, would soon bring biotechnology to the forefront of science in society, and the intimate relationship between the scientific community, the public, and the government would ensue. These debates gained exposure in 1975 at the Asilomar Conference, where Joshua Lederberg was the most outspoken supporter for this emerging field in biotechnology. By as early as 1978, with the synthesis of human insulin, Lederberg's claims would prove valid, and the biotechnology industry grew rapidly. Each new scientific advance became a media event designed to capture public support, and by the 1980s, biotechnology grew into a promising real industry. In 1988, only five proteins from genetically engineered cells had been approved as drugs by the FDA, but this number would skyrocket to over 125 by the end of the 1990s.

The field of genetic engineering remains a heated topic of discussion in today's society with the advent of gene therapy, stem cell research, cloning, and genetically-modified food. While it seems only natural nowadays to link pharmaceutical drugs as solutions to health and societal problems, this relationship of biotechnology serving social needs began centuries ago.

Origins of Biotechnology

Biotechnology arose from the field of zymotechnology, which began as a search for a better understanding of industrial fermentation, particularly beer. Beer was an important industrial, and not just social, commodity. In late 19th century Germany, brewing contributed as much to the gross national product as steel, and taxes on alcohol proved to be significant sources of revenue to the government. In the 1860s, institutes and remunerative consultancies were dedicated to the technology of brewing. The most famous was the private Carlsberg Institute, founded in 1875, which employed Emil Christian Hansen, who pioneered the pure yeast process for the reliable production of consistent beer. Less well known were private consultancies that advised the brewing industry. One of these, the Zymotechnic Institute, was established in Chicago by the German-born chemist John Ewald Siebel.

The heyday and expansion of zymotechnology came in World War I in response to industrial needs to support the war. Max Delbruck grew yeast on an immense scale during the war to meet 60 per cent of Germany's animal feed needs. Compounds of another fermentation product, lactic acid, made up for a lack of hydraulic fluid, glycerol. On the Allied side the Russian chemist, Chaim Weizmann used starch to eliminate Britain's shortage of acetone, a key raw material in explosives, by fermenting maize to acetone. The industrial potential of fermentation was outgrowing its traditional home in brewing, and "zymotechnology" soon gave way to "biotechnology."

With food shortages spreading and resources fading, some dreamed of a new industrial solution. Karl Ereky coined the word "biotechnology" in Hungary during 1917 to describe a technology based on converting raw materials into a more useful product. He built a slaughterhouse for a thousand pigs and also a fattening farm with space for 50,000 pigs, raising over 100,000 pigs a year. The enterprise was enormous, becoming one of the largest and most profitable meat and fat operations in the world. In a book entitled *Biotechnologie,* Ereky further developed a theme that would be reiterated through the 20th century: biotechnology could provide solutions to societal crises, such as food and energy shortages. For Ereky, the term "biotechnologie" indicated the process by which raw materials could be biologically upgraded into socially useful products.

This catchword spread quickly after the First World War,

as "biotechnology" entered German dictionaries and was taken up abroad by business-hungry private consultancies as far away as the United States. In Chicago, for example, the coming of prohibition at the end of World War I encouraged biological industries to create opportunities for new fermentation products, in particular a market for non-alcoholic drinks. Emil Siebel, the son of the founder of the Zymotechnic Institute, broke away from his father's company to establish his own called the "Bureau of Biotechnology," which specifically offered expertise in fermented non-alcoholic drinks.

The belief that the needs of an industrial society could be met by fermenting agricultural waste was an important ingredient of the "chemurgic movement." Fermentation-based processes generated products of ever-growing utility. In the 1940s, penicillin was the most dramatic. While it was discovered in England, it was produced industrially in the US using a deep fermentation process originally developed in Peoria, Illinois. The enormous profits and the public expectations penicillin engendered caused a radical shift in the standing of the pharmaceutical industry. Doctors used the phrase "miracle drug," and the historian of its wartime use, David Adams, has suggested that to the public penicillin represented the perfect health that went together with the car and the dream house of wartime American advertising. In the 1950s, steroids were synthesised using fermentation technology. In particular, cortisone promised the same revolutionary ability to change medicine as penicillin had.

Single-cell Protein

Even greater expectations of biotechnology were raised during the 1960s by a process that grew single-cell protein. When the so-called protein gap threatened world hunger, producing food locally by growing it from waste seemed to offer a solution. It was the possibilities of growing microorganisms on oil that captured the imagination of scientists, policy makers, and commerce. Major companies such as British Petroleum (BP) staked their futures on it. In

1962, BP built a pilot plant at Cap de Lavera in Southern France to publicise its product, Toprina. As there was no well-accepted term to describe the new foods, in 1966 the term "single-cell protein" (SCP) was coined at MIT to provide an acceptable and exciting new title, avoiding the unpleasant connotations of microbial or bacterial. By the late 1970s, however, the cultural climate had completely changed, as the growth in SCP interest had taken place against a shifting economic and cultural scene. First, the price of oil rose catastrophically in 1974, so that its cost per barrel was five times greater than it had been two years earlier. Second, despite continuing hunger around the world, anticipated demand also began to shift from humans to animals.

The programme had begun with the vision of growing food for Third World people, yet the product was instead launched as an animal food for the developed world. The rapidly rising demand for animal feed made that market appear economically more attractive. The ultimate downfall of the SCP project, however, came from public resistance. This was particularly vocal in Japan, where production came closest to fruition. For all their enthusiasm for innovation and traditional interest in microbiologically produced foods, the Japanese were the first to ban the production of single-cell proteins.

The Japanese ultimately were unable to separate the idea of their new "natural" foods from the far from natural connotation of oil. These arguments were made against a background of suspicion of heavy industry in which anxiety over minute traces of petroleum was expressed. Thus, public resistance to an unnatural product led to the end of the SCP project as an attempt to solve world hunger.

In the late 1970s, biotechnology offered another possible solution to a societal crisis. The escalation in the price of oil in 1974 increased the cost of the Western world's energy tenfold. In response, the US government promoted the production of gasohol, gasoline with 10 per cent alcohol added, as an answer to the energy crisis. In 1979, when the Soviet Union sent troops to Afghanistan, the Carter administration cut off its supplies to agricultural produce in retaliation, creating a surplus of agriculture in the US. As a result, fermenting the agricultural surpluses to synthesise fuel seemed to be an economical solution to the shortage of oil threatened by the Iran-Iraq war. Before the new direction could be taken, however, the political wind changed again: the Reagan administration came to power in January 1981 and, with the declining oil prices of the 1980s, ended support for the gasohol industry before it was born.

Biotechnology seemed to be the solution for major social problems, including world hunger and energy crises. In the 1960s, radical measures would be needed to meet world starvation, and biotechnology seemed to provide an answer. However, the solutions proved to be too expensive and socially unacceptable, and solving world hunger through SCP food was dismissed. In the 1970s, the food crisis was succeeded by the energy crisis, and here too, biotechnology seemed to provide an answer. But once again, costs proved prohibitive as oil prices slumped in the 1980s. Thus, in practice, the implications of biotechnology were not fully realised in these situations. But this would soon change with the rise of genetic engineering.

Genetic Engineering

The origins of biotechnology culminated with the birth of genetic engineering. There were two key events that have come to be seen as scientific breakthroughs beginning the era that would unite genetics with biotechnology. One was the 1953 discovery of the structure of DNA, by Watson and Crick, and the other was the 1973 discovery by Cohen and Boyer of a recombinant DNA technique by which a section of DNA was cut from the plasmid of an E. coli bacterium and transferred into the DNA of another. This approach could, in principle, enable bacteria to adopt the genes and produce proteins of other organisms, including humans. Popularly referred to as "genetic engineering," it came to be defined as the basis of new biotechnology.

Genetic engineering proved to be a topic that thrust biotechnology into the public scene, and the interaction between scientists, politicians, and the public defined the work that was accomplished in this area. Technical developments during this time were revolutionary and at times frightening. In December 1967, the first heart transplant by Christian Barnard reminded the public that the physical identity of a person was becoming increasingly problematic. While poetic imagination had always seen the heart at the centre of the soul, now there was the prospect of individuals being defined by other people's hearts. During the same month, Arthur Kornberg announced that he had managed to biochemically replicate a viral gene. "Life had been synthesised," said the head of the National Institutes of Health. Genetic engineering was now on the scientific agenda, as it was becoming possible to identify genetic characteristics with diseases such as beta thalassemia and sickle-cell anaemia.

Responses to scientific achievements were coloured by cultural skepticism. Scientists and their expertise were looked upon with suspicion. In 1968, an immensely popular work, *The Biological Time Bomb,* was written by the British journalist Gordon Rattray Taylor. The author's preface saw Kornberg's discovery of replicating a viral gene as a route to lethal doomsday bugs. The publisher's blurb for the book warned that within ten years, "You may marry a semi-artificial man or woman. The book ended with a chapter called "The Future – If Any." While it is rare for current science to be represented in the movies, in this period of "Star Trek", science fiction and science fact seemed to be converging. "Cloning" became a popular word in the media. Woody Allen satirised the cloning of a person from a nose in his 1973 movie *Sleeper,* and cloning Hitler from surviving cells was the theme of the 1976 novel by Ira Levin, *The Boys from Brazil.*

In response to these public concerns, scientists, industry, and governments increasingly linked the power of recombinant DNA to the immensely practical functions that biotechnology promised. One of the key scientific figures that attempted to highlight the promising aspects of genetic engineering was Joshua Lederberg, a Stanford professor and Nobel laureate. While in the 1960s "genetic engineering" described eugenics and work involving the manipulation of the human genome, Lederberg stressed research that would involve microbes instead. Lederberg emphasised the importance of focusing on curing living people. Lederberg's 1963 paper, "Biological Future of Man" suggested that, while molecular biology might one day make it possible to change the human genotype, "what we have overlooked is euphenics, the engineering of human development." Lederberg constructed the word "euphenics" to emphasise changing the phenotype after conception rather than the genotype which would affect future generations.

With the discovery of recombinant DNA by Cohen and Boyer in 1973, the idea that genetic engineering would have major human and societal consequences was born. In July 1974, a group of eminent molecular biologists headed by Paul Berg wrote to *Science* suggesting that the consequences of this work were so potentially destructive that there should be a pause until its implications had been thought through. This suggestion was explored at a meeting in February 1975 at California's Monterey Peninsula, forever immortalised by the location, Asilomar. Its historic outcome was an unprecedented call for a halt in research until it could be regulated in such a way that the public need not be anxious, and it led to a 16-month moratorium until NIH guidelines were established.

Lederberg was the leading exception in emphasising, as he had for years, the potential benefits. At Asilomar, in an atmosphere favouring control and regulation, he circulated a paper countering the pessimism and fears of misuses with the benefits conferred by successful use. He described "an early chance for a technology of untold importance for diagnostic and therapeutic medicine: the ready production of an unlimited variety of human proteins. Analogous applications may be foreseen in fermentation process for cheaply manufacturing essential nutrients, and in the improvement of microbes for the production of antibiotics and of special industrial chemicals." In June 1976, the 16-month moratorium on research expired with the Director's Advisory Committee (DAC) publication of the NIH guidelines of good practice. They defined the risks of certaih kinds of experiments and the appropriate physical conditions for their pursuit, as well as a list of things too dangerous to perform at all. Moreover, modified organisms were not to be tested outside the confines of a laboratory or allowed into the environment.

Atypical as Lederberg was at Asilomar, his optimistic vision of genetic engineering would soon lead to the development of the biotechnology industry. Over the next two years, as public concern over the dangers of recombinant DNA research grew, so too did interest in its technical and practical applications. Curing genetic diseases remained in the realms of science fiction, but it appeared that producing human proteins would be good business. Insulin, one of the smaller and best understood proteins, had been used in treating diabetes for half a century.

It had been extracted from animals in a chemically slightly different form from the human product. Yet, if one could produce human insulin, one could meet an existing enormous demand with an apparently superior product whose approval would be relatively easy. In the period 1975 to 1977, insulin represented the aspirations for new products that could be made with the new biotechnology. Microbial production of human insulin was finally announced in September 1978 and was commercially produced by a startup company, Genentech.

The radical shift in the connotation of "genetic engineering" from an emphasis on the inherited characteristics of people to the commercial production of proteins and therapeutic drugs was nurtured by Joshua Lederberg. His broad concerns since the 1960s had been stimulated by enthusiasm for science and its potential medical benefits. Countering calls for strict regulation, he expressed a vision of potential utility. Against a belief that new techniques would entail unmentionable and uncontrollable consequences for humanity and the environment, a growing consensus on the economic value of recombinant DNA emerged.

Biotechnology and Industry

With ancestral roots in industrial microbiology that date back centuries, the new biotechnology industry grew rapidly beginning in the mid-1970s. Each new scientific advance became a media event designed to capture investment confidence and public support. Although market expectations and social benefits of new products were frequently overstated, many people were prepared to see genetic engineering as the next great advance in technological progress. By the 1980s, biotechnology characterised a nascent real industry, providing titles for emerging trade organisations such as the Industrial Biotechnology Association.

The main foci of attention after insulin were the potential profit makers in the pharmaceutical industry: human growth hormone and what promised to be a miraculous cure for viral diseases, interferon. Cancer was a central target in the 1970s because increasingly the disease was linked to viruses. By 1980, a new company, Biogen, had produced interferon through recombinant DNA. The emergence of interferon and the possibility of curing cancer raised money in the community for research and increased the enthusiasm of an otherwise uncertain and tentative society. Moreover, to the 1970s plight of cancer was added AIDS in the 1980s, offering an enormous potential market for a successful therapy, and more immediately, a market for diagnostic tests based on monoclonal antibodies. By 1988, only five proteins from genetically engineered cells had been approved as drugs by the FDA: insulin, human growth hormone, hepatitis B vaccine, alpha-interferon, and tissue plasminogen activator (TPa), for lysiş of blood clots. By the end of the 1990s, however, 125 more genetically engineered drugs would be approved.

Genetic engineering also reached the agricultural front as

well. There was tremendous progress since the market introduction of the genetically engineered Flavr Savr tomato in 1994. Ernst and Young reported that in 1998, 30 per cent of the US soybean crop was expected to be from genetically engineered seeds. In 1998, about 30 per cent of the US cotton and corn crops were also expected to be products of genetic engineering.

Genetic engineering in biotechnology stimulated hopes for both therapeutic proteins, drugs and biological organisms themselves, such as seeds, pesticides, engineered yeasts, and modified human cells for treating genetic diseases. From the perspective of its commercial promoters, scientific breakthroughs, industrial commitment, and official support were finally coming together, and biotechnology became a normal part of business. No longer were the proponents for the economic and technological significance of biotechnology the iconoclasts. Their message had finally become accepted and incorporated into the policies of governments and industry.

Global Trends

According to Burrill and Company, an industrial investment bank, over $350 billion has been invested in biotech so far, and global revenues have risen from $23 billion in 2000 to more than $50 billion in 2005. The greatest growth has been in Latin America but all regions of the world have shown strong growth trends.

There has been little innovation in the traditional pharmaceutical industry over the past decade and biopharmaceuticals are now achieving the fastest rates of growth against this background, particularly in breast cancer treatment. Biopharmaceuticals typically treat subsets of the total population with a disease whereas traditional drugs are developed to treat the population as a whole. However, one of the great difficulties with traditional drugs are the toxic side effects the incidence of which can be unpredictable in individual patients.

Many have expressed concerns about the safety, environmental impacts, and social impacts of biotechnology. A book by Michael Mehta (2005) entitled *Biotechnology Unglued: Science, Society and Social Cohesion* (UBC Press) examines the two faces of biotechnology, and provides a series of case-studies on how different applications in biotechnology affect the social cohesiveness of different kinds of communities.

Pharmacogenomics

Pharmacogenomics is the branch of pharmaceutics which deals with the influence of genetic variation on drug response in patients by correlating gene expression or single-nucleotide polymorphisms with a drug's efficacy or toxicity. By doing so, pharmacogenomics aims to develop rational means to optimise drug therapy, with respect to the patients' genotype, to ensure maximum efficacy with minimal adverse effects. Such approaches promise the advent of "personalised medicine", in which drugs and drug combinations are optimised for each individual's unique genetic makeup.

Pharmacogenomics is the whole genome application of pharmacogenetics, which examines the single gene interactions with drugs.

Pharmaceutical Products

Traditional pharmaceutical drugs are small chemicals molecules that treat the symptoms of a disease or illness - one molecule directed at a single target. Biopharmaceuticals are large biological molecules known as proteins and these target the underlying mechanisms and pathways of a malady; it is a relatively young industry. They can deal with targets in humans that are not accessible with traditional medicines. A patient typically is dosed with a small molecule *via* a tablet while a large molecule is typically injected.

Small molecules are manufactured by chemistry but large molecules are created by living cells: for example, - bacteria cells, yeast cell, animal cells.

Modern biotechnology is often associated with the use of genetically altered microorganisms such as *E. coli* or yeast for the production of substances like insulin or antibiotics. It can also refer to transgenic animals or transgenic plants, such as Bt corn. Genetically altered mammalian cells, such as Chinese Hamster Ovary (CHO) cells, are also widely used to manufacture pharmaceuticals. Another promising new biotechnology application is the development of plant-made pharmaceuticals.

Biotechnology is also commonly associated with landmark breakthroughs in new medical therapies to treat diabetes, hepatitis B, hepatitis C, cancers, arthritis, haemophilia, bone fractures, multiple sclerosis, cardiovascular as well as molecular diagnostic devices than can be used to define the patient population. Herceptin, is the first drug approved for use with a matching diagnostic test and is used to treat breast cancer in women whose cancer cells express the protein HER2.

Modern biotechnology can be used to manufacture existing drugs more easily and cheaply. The first genetically engineered products were medicines designed to combat human diseases. To cite one example, in 1978 Genentech joined a gene for insulin and a plasmid vector and put the resulting gene into a bacterium called Escherichia coli. Insulin, widely used for the treatment of diabetes, was previously extracted from sheep and pigs. It was very expensive and often elicited unwanted allergic responses. The resulting genetically engineered bacterium enabled the production of vast quantities of human insulin at low cost.

Since then modern biotechnology has made it possible to produce more easily and cheaply the human growth hormone, clotting factors for hemophiliacs, fertility drugs, erythropoietin and other drugs. Most drugs today are based on about 500 molecular targets. Genomic knowledge of the genes involved in diseases, disease pathways, and drug-response sites are expected to lead to the discovery of thousands more new targets.

Genetic Testing

Genetic testing involves the direct examination of the DNA molecule itself. A scientist scans a patient's DNA sample for mutated sequences.

There are two major types of gene tests. In the first type, a researcher may design short pieces of DNA ("probes") whose sequences are complementary to the mutated sequences. These probes will seek their complement among the base pairs of an individual's genome. If the mutated sequence is present in the patient's genome, the probe will bind to it and flag the mutation. In the second type, a researcher may conduct the gene test by comparing the sequence of DNA bases in a patient's gene to a normal version of the gene.

Genetic Testing can be used to:

- Diagnose a disease.
- Confirm a diagnosis.
- Provide prognostic information about the course of a disease.
- Confirm the existence of a disease in individuals.

With varying degrees of accuracy, predict the risk of future disease in healthy individuals or their progeny.

Genetic Testing is now used for:

- Determining sex.
- Carrier screening, or the identification of unaffected individuals who carry one copy of a gene for a disease that requires two copies for the disease to manifest.
- Prenatal diagnostic screening.
- Newborn screening.
- Presymptomatic testing for predicting adult-onset disorders.
- Presymptomatic testing for estimating the risk of developing adult-onset cancers.

- Confirmational diagnosis of symptomatic individuals.
- Forensic/identity testing.

Some genetic tests are already available, although most of them are used in developed countries. The tests currently available can detect mutations associated with rare genetic disorders like cystic fibrosis, sickle cell anaemia, and Huntington's disease. Recently, tests have been developed to detect mutation for a handful of more complex conditions such as breast, ovarian, and colon cancers.

However, gene tests may not detect every mutation associated with a particular condition because many are as yet undiscovered, and the ones they do detect may present different risks to different people and populations.

Gene Therapy

Gene therapy may be used for treating, or even curing, genetic and acquired diseases like cancer and AIDS by using normal genes to supplement or replace defective genes or to bolster a normal function such as immunity. It can be used to target somatic (i.e., body) or germ (i.e., egg and sperm) cells. In somatic gene therapy, the genome of the recipient is changed, but this change is not passed along to the next generation. In contrast, in germline gene therapy, the egg and sperm cells of the parents are changed for the purpose of passing on the changes to their offspring.

There are basically two ways of implementing a gene therapy treatment:

1. *Ex vivo*, which means "outside the body" – Cells from the patient's blood or bone marrow are removed and grown in the laboratory. They are then exposed to the virus carrying the desired gene. The virus enters the cells, and the desired gene becomes part of the DNA of the cells. The cells are allowed to grow in the laboratory before being returned to the patient by injection into a vein.

2. *In vivo*, which means "inside the body" – No cells are removed from the patient's body. Instead, vectors are used to deliver the desired gene to cells in the patient's body.

Currently, the use of gene therapy is limited. Somatic gene therapy is primarily at the experimental stage. Germline therapy is the subject of much discussion but it is not being actively investigated in larger animals and human beings.

As of June 2001, more than 500 clinical gene-therapy trials involving about 3,500 patients have been identified worldwide. Around 78 per cent of these are in the United States, with Europe having 18 per cent. These trials focus on various types of cancer, although other multigenic diseases are being studied as well. Recently, two children born with severe combined immunodeficiency disorder ("SCID") were reported to have been cured after being given genetically engineered cells.

Gene therapy faces many obstacles before it can become a practical approach for treating disease. At least four of these obstacles are as follows:

1. *Gene Delivery Tools*: Genes are inserted into the body using gene carriers called vectors. The most common vectors now are viruses, which have evolved a way of encapsulating and delivering their genes to human cells in a pathogenic manner. Scientists manipulate the genome of the virus by removing the disease-causing genes and inserting the therapeutic genes. However, while viruses are effective, they can introduce problems like toxicity, immune and inflammatory responses, and gene control and targeting issues.
2. *Limited knowledge of the Functions of Genes*: Scientists currently know the functions of only a few genes. Hence, gene therapy can address only some genes that cause a particular disease. Worse, it is not known exactly whether genes have more than one function, which creates uncertainty as to whether replacing such genes is indeed desirable.

3. *Multigene Disorders and Effect of Environment*: Most genetic disorders involve more than one gene. Moreover, most diseases involve the interaction of several genes and the environment. For example, many people with cancer not only inherit the disease gene for the disorder, but may have also failed to inherit specific tumour suppressor genes. Diet, exercise, smoking and other environmental factors may have also contributed to their disease.
4. *High Costs*: Since gene therapy is relatively new and at an experimental stage, it is an expensive treatment to undertake. This explains why current studies are focused on illnesses commonly found in developed countries, where more people can afford to pay for treatment. It may take decades before developing countries can take advantage of this technology.

Human Genome Project

The Human Genome Project is an initiative of the US Department of Energy ("DOE") that aims to generate a high-quality reference sequence for the entire human genome and identify all the human genes.

The DOE and its predecessor agencies were assigned by the US Congress to develop new energy resources and technologies and to pursue a deeper understanding of potential health and environmental risks posed by their production and use. In 1986, the DOE announced its Human Genome Initiative. Shortly thereafter, the DOE and National Institutes of Health developed a plan for a joint Human Genome Project ("HGP"), which officially began in 1990.

The HGP was originally planned to last 15 years. However, rapid technological advances and worldwide participation have accelerated the expected completion date to 2003. In June 2000, scientists announced the generation of a working draft sequence of the entire human genome. The draft provides a road map to an estimated 90 per cent of genes on every human chromosome. Already it has enabled gene hunters to pinpoint genes associated with more than 30 disorders.

Cloning

Human cloning is one of the techniques of modern biotechnology. It involves the removal of the nucleus from one cell and its placement in an unfertilized egg cell whose nucleus has either been deactivated or removed.

There are Three Types of Cloning:

1. *Reproductive Cloning:* After a few divisions, the egg cell is placed into a uterus where it is allowed to develop into a fetus that is genetically identical to the donor of the original nucleus.
2. *Therapeutic Cloning:* The egg is placed into a Petri dish where it develops into embryonic stem cells, which have shown potentials for treating several ailments.

In February 1997, cloning became the focus of media attention when Ian Wilmut and his colleagues at the Roslin Institute announced the successful cloning of a sheep, named Dolly, from the mammary glands of an adult female. The cloning of Dolly made it apparent to many that the techniques used to produce her could someday be used to clone human beings. This stirred a lot of controversy because of its ethical implications.

Concerns Regarding the use of Modern Biotechnology Techniques in Medicine

Several issues have been raised regarding the use of modern biotechnology in the medical sector. Many of these issues are similar to those facing any new technology that is viewed as powerful and far-reaching. Some of these issues are:

1. *Absence of Cure:* There is still a lack of effective treatment or preventive measures for many diseases and conditions now being diagnosed or predicted using gene tests. Thus, revealing information about risk of

a future disease that has no existing cure presents an ethical dilemma for medical practitioners.

2. *Ownership and Control of Genetic Information:* Who will own and control genetic information, or information about genes, gene products, or inherited characteristics derived from an individual or a group of people like indigenous communities. At the macro level, there is a possibility of a genetic divide, with developing countries that do not have access to medical applications of biotechnology being deprived of benefits accruing from products derived from genes obtained from their own people. Moreover, genetic information can pose a risk for minority population groups as it can lead to group stigmatisation.

At the individual level, the absence of privacy and anti-discrimination legal protections in most countries can lead to discrimination in employment or insurance or other misuse of personal genetic information. This raises questions like, is genetic privacy different from medical privacy?

3. *Reproductive Issues:* These include the use of genetic information in reproductive decision-making and the possibility of genetically altering reproductive cells that may be passed on to future generations. For example, germline therapy forever changes the genetic makeup of an individual's descendants. Thus, any error in technology or judgment may have far-reaching consequences. Ethical issues like designer babies and human cloning have also given rise to controversies between and among scientists and bioethicists, especially in the light of past abuses with eugenics.
4. *Clinical Issues:* These centre on the capabilities and limitations of doctors and other health-service providers, people identified with genetic conditions, and the general public in dealing with genetic information. For instance, how should the public be prepared to make informed choices based on the results

of genetic tests? How will genetic tests be evaluated and regulated for accuracy, reliability, and usefulness?

5. *Effects on Social Institutions:* Genetic tests reveal information about individuals and their families. Thus, test results can affect the dynamics within social institutions, particularly the family.
6. Conceptual and philosophical implications regarding human responsibility, free will vis-a-vis genetic determinism, and the concepts of health and disease. Do genes influence human behaviour? If so, does genetic testing mean controlling human behaviour? What is considered acceptable diversity? What is normal and what is a disability or disorder, and who decides these matters? Are disabilities diseases that need to be cured or prevented? Where should the line between medical treatment and enhancement be drawn; moreover, what considerations make that delineation important and how do such considerations bear on public policy and personal choice? Who will have access to gene therapy?

Role in Agriculture

There are many applications of biotechnology in agriculture.

One is improved yield from crops. Using the techniques of modern biotechnology, one or two genes may be transferred to a highly developed crop variety to impart a new character that would increase its yield. However, while increase in crop yield is the most obvious application of modern biotechnology in agriculture, it is also the most difficult one. Current genetic engineering techniques work best for effects that are controlled by a single gene. Many of the genetic characteristics associated with yield (e.g., enhanced growth) are controlled by a large number of genes, each of which has a minimal effect on the overall yield. There is, therefore, much scientific work to be done in this area.

Another is the reduced vulnerability of crops to environmental stresses. Crops containing genes that will enable them to withstand biotic and abiotic stresses may be developed. For example, drought and excessively salty soil are the two most important limiting factors in crop productivity. Biotechnologists are studying plants that can cope with these extreme conditions in the hope of finding the genes that enable them to do so and eventually transferring these genes to the more desirable crops. One of the latest developments is the identification of a plant gene, At-DBF2, from thale cress, a tiny weed that is often used for plant research because it is very easy to grow and its genetic code is well mapped out. When this gene was inserted into tomato and tobacco cells, the cells were able to withstand environmental stresses like salt, drought, cold and heat, far more than ordinary cells. If these preliminary results prove successful in larger trials, then At-DBF2 genes can help in engineering crops that can better withstand harsh environments.

Researchers have also created transgenic rice plants that are resistant to rice yellow mottle virus (RYMV). In Africa, this virus destroys majority of the rice crops and makes the surviving plants more susceptible to fungal infections.

Increased nutritional qualities of food crops. Proteins in foods may be modified to increase their nutritional qualities. Proteins in legumes and cereals may be transformed to provide the amino acids needed by human beings for a balanced diet. A good example is the work of Professors Ingo Potrykus and Peter Beyer on the so-called Goldenrice.

Improved taste, texture or appearance of food. Modern biotechnology can be used to slow down the process of spoilage so that fruit can ripen longer on the plant and then be transported to the consumer with a still reasonable shelf life. This improves the taste, texture and appearance of the fruit. More importantly, it could expand the market for farmers in developing countries due to the reduction in spoilage.

The first genetically modified food product was a tomato which was transformed to delay its ripening. Researchers in Indonesia, Malaysia, Thailand, Philippines and Vietnam are currently working on delayed-ripening papaya in collaboration with the University of Nottingham and Zeneca.

Most of the current commercial applications of modern biotechnology in agriculture are on reducing the dependence of farmers on agrochemicals. For example, Bacillus thuringiensis (Bt) is a soil bacterium that produces a protein with insecticidal qualities. Traditionally, a fermentation process has been used to produce an insecticidal spray from these bacteria. In this form, the Bt toxin occurs as an inactive protoxin, which requires digestion by an insect to be effective. There are several Bt toxins and each one is specific to certain target insects. Crop plants have now been engineered to contain and express the genes for Bt toxin, which they produce in its active form. When a susceptible insect ingests the transgenic crop cultivar expressing the Bt protein, it stops feeding and soon thereafter dies as a result of the Bt toxin binding to its gut wall. Bt corn is now commercially available in a number of countries to control corn borer (a lepidopteran insect), which is otherwise controlled by spraying (a more difficult process).

Crops have also been genetically engineered to acquire tolerance to broad-spectrum herbicide. The lack of cost-effective herbicides with broad-spectrum activity and no crop injury was a consistent limitation in crop weed management. Multiple applications of numerous herbicides were routinely used to control a wide range of weed species detrimental to agronomic crops. Weed management tended to rely on pre-emergence — that is, herbicide applications were sprayed in response to expected weed infestations rather than in response to actual weeds present. Mechanical cultivation and hand weeding were often necessary to control weeds not controlled by herbicide applications. The introduction of herbicide tolerant crops has the potential of reducing the number of herbicide active ingredients used for weed management, reducing the number of herbicide applications made during a season, and increasing yield due to improved weed management and less crop injury. Transgenic crops that express tolerance to glyphosphate, glufosinate and bromoxynil have been developed. These herbicides can now be sprayed on transgenic crops without inflicting damage on the crops while killing nearby weeds.

From 1996 to 2001, herbicide tolerance was the most dominant trait introduced to commercially available transgenic crops, followed by insect resistance. In 2001, herbicide tolerance deployed in soybean, corn and cotton accounted for 77 per cent of the 62.6 million hectares planted to transgenic crops; Bt crops accounted for 15 per cent; and stacked genes for herbicide tolerance and insect resistance used in both cotton and corn accounted for 8 per cent.

Modern biotechnology is increasingly being applied for novel uses other than food. For example, oilseed is at present used mainly for margarine and other food oils, but it can be modified to produce fatty acids for detergents, substitute fuels and petrochemicals. Banana trees and tomato plants have also been genetically engineered to produce vaccines in their fruit. If future clinical trials prove successful, the advantages of edible vaccines would be enormous, especially for developing countries. The transgenic plants may be grown locally and cheaply. Homegrown vaccines would also avoid logistical and economic problems posed by having to transport traditional preparations over long distances and keeping them cold while in transit. And since they are edible, they will not need syringes, which are not only an additional expense in the traditional vaccine preparations but also a source of infections if contaminated.

Biological Engineering

Bioengineering: Biotechnological engineering or biological engineering is a branch of engineering that focuses on biotechnologies and biological science. It includes different disciplines such as biochemical engineering, biomedical engineering, bioprocess engineering, biosystem engineering and so on. Because of the novelty of the field, the definition of a bioengineer is still undefined. However, in general it is an integrated approach of fundamental biological sciences and traditional engineering principles.

Bioengineers are often employed to scale up bio processes from the laboratory scale to the manufacturing scale. Moreover, as with most engineers, they often deal with management, economic and legal issues. Since patents and regulation (e.g. FDA regulation in the US) are very important issues for biotech enterprises, bioengineers are often required to have knowledge related to these issues.

The increasing number of biotech enterprises is likely to create a need for bioengineers in the years to come. Many universities throughout the world are now providing programmes in bioengineering and biotechnology (as independent programmes or speciality programmes within more established engineering fields).

Agrobacterium

Agrobacterium is a genus of bacteria that causes tumours in plants. *Agrobacterium tumefaciens* is the most commonly studied species in this genus. *Agrobacterium* is well known for its ability to transfer DNA between itself and plants, and for this reason it has become an important tool for plant improvement by genetic engineering.

Taxonomic Note: The *Agrobacterium* genus is quite heterogeneous. Recent taxonomic studies have reclassified all of the *Agrobacterium* species into new genera, such as *Ruegeria*, *Pseudorhodobacter* and *Stappia*, but most species have been reclassified as *Rhizobium* species.

Agrobacterium as a Plant Pathogen

A. tumefaciens causes crown-gall disease in plants. The disease is characterised by a tumour like growth or gall on the infected plant, often at the junction between the root and the shoot. Tumours are incited by the transfer of a DNA (T-DNA) segment from the bacterial tumour-inducing (Ti) plasmid. The closely related species, *A. rhizogenes*, induces root tumours, and carries the distinct Ri (root-inducing) plasmid. Although the taxonomy of *Agrobacterium* is currently under revision, it can be generalised that 3 biovars exist within the genus, *A. tumefaciens* or biovar 1, *A. rhizogenes* or biovar 2, and *A. vitis* or biovar 3. Strains within biovars 1 and 2 are known to be able to harbour either a Ti or Ri-plasmid, whilst strains of biovar 3, generally restricted to grapevines, can harbour a Ti-plasmid. Non-Agrobacterium strains have been isolated from environmental samples which harbour a Ri-plasmid whilst laboratory studies have shown that non-Agrobacterium strains can also harbour a Ti-plasmid. Many environmental strains of *Agrobacterium* do not possess either a Ti or Ri-plasmid. These strains are avirulent.

The plasmid T-DNA is integrated semi-randomly into the genome of the host cell (Francis and Spiker, 2005. Plant Journal. 41(3): 464.), and the virulence (vir) genes on the T-DNA are expressed, causing the formation of a gall. The T-DNA carries genes for the biosynthetic enzymes for the production of unusual amino acids, typically octopine or nopaline. It also carries genes for the biosyntheis of the plant hormones, auxin and cytokinins. By altering the hormone balance in the plant cell, the division of those cells cannot be controlled by the plant, and tumours form. The ratio of auxin to cytokinin produced by the tumour genes determines the morphology of the tumour (root-like, disorganised or shoot-like).

Agrobacterium in Biotechnology

The ability of *Agrobacterium* to transfer genes to plants has been exploited for genetic engineering for plant improvement. A modified Ti or Ri plasmid can be used. The plasmid is 'disarmed' by deletion of the tumour inducing genes, the only essential parts of the T-DNA are its two small (25 base pair) border repeats, at least one of which is needed for plant transformation. Marc Van Montagu and Jozef Schell at the University of Ghent (Belgium) discovered the gene transfer mechanism between Agrobacterium and plants, which resulted in the development of methods to alter Agrobacterium into an efficient delivery system for gene engineering in plants. A team of researchers led by Dr. Mary-Dell Chilton were the first to demonstrate that the virulence genes could be removed without adversely affecting the ability of *Agrobacterium* to insert its own DNA into the plant genome (1983).

The genes to be introduced into the plant are cloned into the T-DNA region of the disarmed plasmid, together with a selectable marker (such as antibiotic resistance) to enable selection for plants that have been successfully transformed. Plants are grown on media containing antibiotic following transformation, and those that do not have the T-DNA integrated into their genome will die. An alternative method is agroinfiltration.

Transformation with *Agrobacterium* can be achieved in two ways. Protoplasts, or leaf-discs can be incubated with the *Agrobacterium* and whole plants regenerated using plant tissue culture. A common transformation protocol for *Arabidopsis* is the floral-dip method: the flowers are dipped in an *Agrobacterium* culture, and the bacterium transforms the germline cells that make the female gametes. The seeds can then be screened for antibiotic resistance (or another marker of interest), and plants that have not integrated the plasmid DNA will die.

Agrobacterium does not infect all plant species, but there are several other effective techniques for plant transformation including the gene gun.

Affymetrix

Affymetrix is a manufacturer of DNA microarrays, based in Santa Clara, California, United States. The company was co-founded by Dr. Stephen Fodor in 1992. The company was begun as a unit in Affymax N.V. in 1991 by Fodor's group, which had in the late 1980's developed methods for fabricating DNA microarrays, called "GeneChips" according to the Affymetrix trademark, using semi-conductor manufacturing techniques. The company's first product, an HIV genotyping

GeneChip, was introduced in 1994 and the company went public in 1996. As a result of their pioneering work and the ensuing popularity of microrarray products, Affymetrix derives significant benefit from its patent portfolio in this area.

Acquisitions have included Genetic Microsystem for slide-based microarrays and scanners and Neomorphic for bioinformatics. In 2000 Perlegen Sciences was spun out to focus on wafer-scale genomics for massive data creation and collection required for characterising population variance of genomic markers and expression for the drug discovery process.

Early on Affymetrix formed a partnership with Hewlett-Packard Chemical Analysis, now Agilent Life Sciences. That relationship soured when Affymetrix acquired its own world-class instrument development team and now the two companies are competitors. Today, using the instrument development and manufacturing teams based in Bedford, MA that evolved from the acquisition of Genetic Microsystems, Affymetrix designs and manufactures its own genomics instrumentation systems as well as performing advanced research in creating new technologies to be used in future genomic instrumentation systems.

Description of Product

Affymetrix makes quartz chips for analysis of DNA Microarrays. These chips are sold under the trademarked name GeneChip. Affymetrix's GeneChips assist researchers in quickly scanning for the presence of particular genes in a biological sample. Within this area, Affymetrix is focused on oligonucleotide microarrays. These microarrays are used to determine which genes exist in a sample by detecting specific pieces of mRNA. A single chip can be used to do thousands of experiments in parallel. Chips can be used only once.

Affymetrix sells both mass produced GeneChips intended to match scientifically important parts of human and other animal genomes. It also operates the Affymetrix Made-to-Order Programme which makes custom chips available to researchers.

Affymetrix manufactures its GeneChips using photolithography. Glass chips are relatively expensive; list prices range from $50-$500 depending on the capabilities of the single-use chip.

Competitors in the DNA Microarray business include Illumina, GE Health care, Ocimum Biosolutions, Applied Biosystems, Beckman Coulter, Eppendorf Biochip Systems, and Agilent. There are also various inexpensive plastic-based technologies under development in small companies and laboratories around the world. It has been widely speculated that mass-produced plastic chips can be produced at lower prices than Affymetrix's quartz chips.

The company also manufactures machinery for high speed analysis of biological samples. List prices for the newest DNA arrayers, which prepare and place samples onto chips, are around $250,000 each. Scanners which read the results from the chip list from $150,000 for complete starter systems to $325,000 for high-throughput systems. Substantially lower retail prices have been seen in the market.

The business model of Affymetrix is based on these sources for revenue: consumables; instruments; subscription, service, license; royalty; and sales to Perlegen Sciences. Affymetrix currently controls 82 per cent of the DNA microarray market in terms of sales.

Affymetrix has established a licensing programme to make its intellectual property accessible to stimulate the broad commercialisation of genome analysis technologies. Licensing fees have offset weak product sales in Q3-2006. They have several collaboration relationships with other companies that utilise their patented GeneChip technology.

Currently Affymetrix is fighting a patent infringement lawsuit against Illumina (company), alleging the infringement of six Affymetrix patents. A pre-trial hearing favours Affymetrix, but the trial is not until May 2007.

Functioning Biotechnology

All organisms are made up of cells that are programmed by the same basic genetic material, called DNA (deoxyribonucleic acid). Each unit of DNA is made up of a combination of the following nucleotides — adenine (A), guanine (G), thymine (T), and cytosine (D) — as well as a sugar and a phosphate. These nucleotides pair up into strands that twist together into a spiral structure call a "double helix." This double helix is DNA. Segments of the DNA tell individual cells how to produce specific proteins. These segments are genes. It is the presence or absence of the specific protein that gives an organism a trait or characteristic. More than 10,000 different genes are found in most plant and animal species. This total set of genes for an organism is organised into chromosomes within the cell nucleus. The process by which a multicellular organism develops from a single cell through an embryo stage into an adult is ultimately controlled by the genetic information of the cell, as well as interaction of genes and gene products with environmental factors.

When cells reproduce, the DNA strands of the double helix separate. Because nucleotide A always pairs with T and G always pairs with C, each DNA strand serves as a precise blueprint for a specific protein. Except for mutations or mistakes in the replication process, a single cell is equipped with the information to replicate into millions of identical cells. Because all organisms are made up of the same type of genetic material (nucleotides A, T, G, and C), biotechnologists use enzymes to cut and remove DNA segments from one organism and recombine it with DNA in another organism. This is called recombinant DNA (rDNA) technology, and it is one of the basic tools of modern biotechnology. rDNA technology is the laboratory manipulation of DNA in which DNA, or fragments of DNA from different sources, are cut and recombined using enzymes. This recombinant DNA is then inserted into a living organism. rDNA technology is usually used synonymously with genetic engineering. rDNA technology allows researchers to move genetic information between unrelated organisms to produce desired products or characteristics or to eliminate undesirable characteristics.

Genetic engineering is the technique of removing, modifying or adding genes to a DNA molecule in order to change the information it contains. By changing this information, genetic engineering changes the type or amount of proteins an organism is capable of producing. Genetic engineering is used in the production of drugs, human gene therapy, and the development of improved plants. For example, an "insect protection" gene (Bt) has been inserted into several crops - corn, cotton, and potatoes - to give farmers new tools for integrated pest management. Bt corn is resistant to European corn borer. This inherent resistance thus reduces a farmers pesticide use for controlling European corn borer, and in turn requires less chemicals and potentially provides higher yielding Agricultural Biotechnology.

Although major genetic improvements have been made in crops, progress in conventional breeding programmes has been slow. In fact, most crops grown in the US produce less than their full genetic potential. These shortfalls in yield are due to the inability of crops to tolerate or adapt to environmental stresses, pests, and diseases. For example, some of the world's highest yields of potatoes are in Idaho under irrigation, but in 1993 both quality and yield were severely reduced because of cold, wet weather and widespread frost damage during June. Some of the world's best bread wheats and malting barleys are produced in the north-central states, but in 1993 the disease Fusarium caused an estimated $1 billion in damage.

Scientists have the ability to insert genes that give biological defence against diseases and insects, thus reducing the need for chemical pesticides, and they will soon be able to convey genetic traits that enable crops to better withstand harsh conditions, such as drought. The International Laboratory for Tropical Agricultural Biotechnology (ILTAB) is developing transformation techniques and applications for control of diseases caused by plant viruses in tropical plants such as rice, cassava and tomato. In 1995, ILTAB reported the first transfer through biotechnology of a resistance gene from a wild species of rice to a susceptible cultivated rice variety. The transferred gene expressed resistance to Xanthomonas oryzae, a bacterium which can destroy the crop through disease. The resistant gene was transferred into susceptible rice varieties that are cultivated on more than 24 million hectares around the world.

Benefits can also be seen in the environment, where insect-protected biotech crops reduce the need for chemical pesticide use. Insect-protected crops allow for less potential exposure of farmers and groundwater to chemical residues, while providing farmers with season-long control. Also by reducing the need for pest control, impacts and resources spent on the land are less, thereby preserving the topsoil.

Major advances also have been made through conventional breeding and selection of livestock, but significant gains can still be made by using biotechnology. Currently, farmers in the U.S spend $17 billion on animal health. Diseases such as hog cholera and pests such as screwworm have been eradicated. Uses of biotechnology in animal production include development of vaccines to protect animals from disease, production of several calves from one embryo (cloning), increase of animal growth rate, and rapid disease detection.

Modern biotechnology has offered opportunities to produce more nutritious and better tasting foods, higher crop yields and plants that are naturally protected from disease and insects. Modern biotechnology allows for the transfer of only one or a few desirable genes, thereby permitting scientists to develop crops with specific beneficial traits and reduce undesirable traits. Traditional biotechnology such as cross-pollination in corn produces numerous, non-selective changes. Genetic modifications have produced fruits that can ripen on the vine for better taste, yet have longer shelf lives through delayed pectin degradation. Tomatoes and other produce containing increased levels of certain nutrients, such as vitamin C, vitamin

E, and or beta-carotene, and help protect against the risk of chronic diseases, such as some cancers and heart disease. Similarly introducing genes that increase available iron levels in rice threefold is a potential remedy for iron deficiency, a condition that effects more than two billion people and causes anaemia in about half that number. Most of the today's hard cheese products are made with a biotech enzyme called chymosin. This is produced by genetically engineered bacteria which is considered more purer and plentiful than it's naturally occurring counterpart, rennet, which is derived from calf stomach tissue.

In 1992, Monsanto Company successfully inserted a gene from a bacterium into the Russet Burbank potato. This gene increases the starch content of the potato. Higher starch content reduces oil absorption during frying, thereby lowering the cost of processing French fries and chips and reducing the fat content in the finished product. This product is still awaiting final development and approval.

Modern biotechnology offers effective techniques to address food safety concerns. Biotechnical methods may be used to decrease the time necessary to detect foodborne pathogens, toxins, and chemical contaminants, as well as to increase detection sensitivity. Enzymes, antibodies, and microorganisms produced using rDNA techniques are being used to monitor food production and processing systems for quality control.

Biotechnology can compress the time frame required to translate fundamental discoveries into applications. This is done by controlling which genes are altered in an organised fashion. For example, a known gene sequence from a corn plant can be altered to improve yield, increase drought tolerance, and produce insect resistance (Bt) in one generation. Conventional breeding techniques would take several years. Conventional breeding techniques would require that a field of corn is grown and each trait is selected from individual stalks of corn. The ears of corn from selected stalks with each desired trait (e.g., drought tolerance and yield performance) would then be grown and combined (cross-pollinated). Their offspring (hybrid) would be further selected for the desired result (a high performing corn with drought tolerance). With improved technology and knowledge about agricultural organisms, processes, and ecosystems, opportunities will emerge to produce new and improved agricultural products in an environmentally sound manner.

In summary, modern biotechnology offers opportunities to improve product quality, nutritional content, and economic benefits. The genetic makeup of plants and animals can be modified by either insertion of new useful genes or removal of unwanted ones. Biotechnology is changing the way plants and animals are grown, boosting their value to growers, processors, and consumers.

Industrial Biotechnology

Industrial biotechnology applies the techniques of modern molecular biology to improve the efficiency and reduce the environmental impacts of industrial processes like textile, paper and pulp, and chemical manufacturing. For example, industrial biotechnology companies develop biocatalysts, such as enzymes, to synthesise chemicals. Enzymes are proteins produced by all organisms. Using biotechnology, the desired enzyme can be manufactured in commercial quantities.

Commodity chemicals (e.g., polymer-grade acrylamide) and speciality chemicals can be produced using biotech applications. Traditional chemical synthesis involves large amounts of energy and often-undesirable products, such as HCl. Using biocatalysts, the same chemicals can be produced more economical and more environmental-friendly. An example would be the substitution of protease in detergents for other cleaning compounds. Detergent proteases, which remove protein impurities, are essential components of modern detergents. They are used to break down protein, starch, and fatty acids present on items being washed. Protease production results in a biomass that in turn yields a useful by-product-an organic fertilizer. Biotechnology is also used in the textile industry for the finishing of fabrics and garments. Biotechnology also produces biotech-derived cotton that is warmer, stronger, has improved dye uptake and retention, enhanced absorbency, and wrinkle- and shrink-resistance.

Some agricultural crops, such as corn, can be used in place of petroleum to produce chemicals. The crop's sugar can be fermented to acid, which can be then used as an intermediate to produce other chemical feedstocks for various products. It has been projected that 30 per cent of the world's chemical and fuel needs could be supplied by such renewable resources. It has been demonstrated, at test scale, that biopulping reduces the electrical energy required for wood pulping process by 30 per cent.

Significance of Environmental Biotechnology

Environmental biotechnology is the used in waste treatment and pollution prevention. Environmental biotechnology can more efficiently clean up many wastes than conventional methods and greatly reduce our dependence on methods for land-based disposal.

Every organism ingests nutrients to live and produces by-products as a result. Different organisms need different types of nutrients. Some bacteria thrive on the chemical components of waste products. Environmental engineers use bioremediation, the broadest application of environmental biotechnology, in two basic ways. They introduce nutrients to stimulate the activity of bacteria already present in the soil at a waste site, or add new bacteria to the soil. The bacteria digest the waste at the site and turn it into harmless by-products. After the bacteria consume the waste materials, they die off or return to their normal population levels in the environment.

Bioremediation, is an area of increasing interest. Through application of biotechnical methods, enzyme bioreactors are being developed that will pretreat some industrial waste and food waste components and allow their removal through the sewage system rather than through solid waste disposal mechanisms. Waste can also be converted to biofuel to run generators. Microbes can be induced to produce enzymes needed to convert plant and vegetable materials into building blocks for biodegradable plastics.

In some cases, the by-products of the pollution-fighting microorganisms are themselves useful. For example, methane can be derived from a form of bacteria that degrades sulphur liquor, a waste product of paper manufacturing. This methane can then be used as a fuel or in other industrial processes.

Human Applications

Biotechnical methods are now used to produce many proteins for pharmaceutical and other specialised purposes. A harmless strain of Escherichia coli bacteria, given a copy of the gene for human insulin, can make insulin. As these genetically modified (GM) bacterial cells, they produce human insulin, which can be purified and used to treat diabetes in humans. Microorganisms can also be modified to produce digestive enzymes. In the future, these microorganisms could be colonised in the intestinal tract of persons with digestive enzyme insufficiencies. Products of modern biotechnology include artificial blood vessels from collagen tubes coated with a layer of the anti-coagulant heparin.

Gene therapy—altering DNA within cells in an organism to treat or cure a disease—is one of the most promising areas of biotechnology research. New genetic therapies are being developed to treat diseases such as cystic fibrosis, AIDS and cancer.

DNA fingerprinting is the process of cross matching two strands of DNA. In criminal investigations, DNA from samples of hair, bodily fluids or skin at a crime scene are compared with those obtained from the suspects. In practice, it has become one of the most powerful and widely known applications of biotechnology today. Another process, polymerase chain reaction (PCR), is also being used to more quickly and accurately identify the presence of infections such as AIDS, Lyme disease and Chlamydia.

Paternity determination is possible because a child's DNA pattern is inherited, half from the mother and half from the father. To establish paternity, DNA fingerprints of the mother, child and the alleged father are compared. The matching sequences of the mother and the child are eliminated from the child's DNA fingerprint; what remains comes from the biological father. These segments are then compared for a match with the DNA fingerprint of the alleged father.

DNA testing is also used on human fossils to determine how closely related fossil samples are from different geographic locations and geologic areas. The results shed light on the history of human evolution and the manner in which human ancestors settled different parts of the world.

Biotechnology for the New Century

Experts in United States anticipate the world's population in 2050 to be approximately 8.7 billion persons. The world's population is growing, but its surface area is not. Compounding the effects of population growth is the fact that most of the earth's ideal farming land is already being utilised. To avoid damaging environmentally sensitive areas, such as rain forests, we need to increase crop yields for land currently in use. By increasing crop yields, through the use of biotechnology the constant need to clear more land for growing food is reduced.

Countries in Asia, Africa, and elsewhere are grappling with how to continue feeding a growing population. They are also trying to benefit more from their existing resources. Biotechnology holds the key to increasing the yield of staple crops by allowing farmers to reap bigger harvests from currently cultivated land, while preserving the land's ability to support continued farming.

Malnutrition in underdeveloped countries is also being combated with biotechnology. The Rockefeller Foundation is sponsoring research on "golden rice", a crop designed to improve nutrition in the developing world. Rice breeders are using biotechnology to build Vitamin A into the rice. Vitamin A deficiency is a common problem in poor countries. A second phase of the project will increase the iron content in rice to combat anaemia, which is widespread problem among women and children in underdeveloped countries. Golden rice, expected to be for sale in Asia in less than five years, will offer dramatic improvements in nutrition and health for millions of people, with little additional costs to consumers.

Similar initiatives using genetic manipulation are aimed at making crops more productive by reducing their dependence on pesticides, fertilizers and irrigation, or by increasing their resistance to plant diseases.

Increased crop yield, greater flexibility in growing environments, less use of chemical pesticides and improved nutritional content make agricultural biotechnology, quite literally, the future of the world's food supply.

Concerns about Biotechnology

As biotechnology has become widely used, questions and concerns have also been raised. The most vocal opposition has come from European countries. One of the main areas of concern is the safety of genetically engineered food.

In assessing the benefits and risks involved in the use of modern biotechnology, there are a series of issues to be addressed so that informed decisions can be made. In making value judgments about risks and benefits in the use of biotechnology, it is important to distinguish between technology-inherent risks and technology-transcending risks. The former includes assessing any risks associated with food safety and the behaviour of a biotechnology-based product in the environment. The latter involve the political and social context in which the technology is used, including how these uses may benefit or harm the interests of different groups in society.

The health effects of foods grown from genetically

engineered crop depend on the composition of the food itself. Any new product may have either beneficial or occasional harmful effects on human health. For example, a biotech-derived food with a higher content of digestible iron is likely to have a positive effect if consumed by iron-deficient individuals. Alternatively, the transfer of genes from one species to another may also transfer the risk for exposure to allergens. These risks are systematically evaluated by FDA and identified prior to commercialisation. Individuals allergic to certain nuts, for example, need to know if genes conveying this trait are transferred to other foods such as soybeans. Labelling would be required if such crops were available to consumers.

Among the potential ecological risks identified are increased weediness, due to cross- pollination from genetically modified crops spreads to other plants in nearby fields. This may allow the spread of traits such as herbicide-resistance to non-target plants that could potentially develop into weeds. This ecological risk is assessed when deciding if a plant with a given trait should be released into a particular environment, and if so, under what conditions.

Other potential ecological risks stem from the use of genetically modified corn and cotton with insecticidal genes from Bacillus thuringiensis (Bt genes). This may lead to the development of resistance to Bt in insect populations exposed to the biotech-derived crop. There also may be risks to non-target species, such as birds and butterflies, from the plants with Bt genes. The monitoring of these effects of new crops in the environment and implementation of effective risk management approaches is an essential component of further research. It is also important to keep all risks in perspective by comparing the products of biotechnology and conventional agriculture.

The reduction of biodiversity would represent a technology-transcending risk. Reduced biological diversity due to destruction of tropical forests, conversion of land to agriculture, overfishing, and the other practices to feed a growing world population is a significant loss far more than any potential loss of biodiversity due to biotech-derived crop varieties. Improved governance and international support are necessary to limit loss of biodiversity.

What we know from our understanding of science and more than a decade of experience with biotech-derived plants is the following: There is no evidence that genetic transfers between unrelated organisms pose human health concerns that are different from those encountered with any new plant or animal variety. The risks associated with biotechnology are the same as those associated with plants and microbes developed by conventional methods.

Consumer and Food Industry Perspectives

Survey research over the past decade shows that biotechnology is not likely to become an important issue for most American consumers. Consumers find biotechnology acceptable when they believe it offers benefits and it is safe. Surveys have consistently found that a majority of American consumers are willing to buy insect-protected food crops developed through biotechnology that use fewer chemical pesticides, as well as more nutritious foods. American consumers also appreciate the role that biotechnology can play in feeding the world. Research shows that European consumers are much less supportive of all biotechnology applications.

Surveys since 1992 show that relatively few US consumers have heard or read much about biotechnology. News about the cloned sheep pushed awareness to 50 per cent in March 1997. Surveys in the first three months of 2000 show that awareness has fallen back to just over one-third in the United States. Such trends reflect the fact that most people get their information about biotechnology from the media. Unfortunately, many consumers also do not understand some fundamental principles of biology. European consumer awareness is somewhat higher, but knowledge is still low.

Media coverage in the United States has generally been

balanced (which helps account for our relatively high levels of acceptance). This is in sharp contrast to the European media, which have played upon fear of the unknown. The European media have also tended to accept opponents' claims without question. Another issue is that many people no longer have a connection to agriculture. In fact, research has shown that many consumers are unaware that all foods are derived from plants or animals that already have been genetically modified through traditional (but imprecise) breeding methods.

American consumers look to health professionals and scientific experts for credible information, but place relatively little trust in the activists who oppose biotechnology. Research shows that acceptance increases significantly when American consumers learn that organisations such as the National Academy of Sciences and the US Food and Drug Administration have determined that biotech-derived foods are safe. In contrast, European consumers express the most trust in those groups that oppose biotechnology.

They have much less confidence in government, industry, or even scientists. American culture is more supportive and rewarding of new technology. Europeans tend to view food differently from US consumers. In fact, some Europeans reject all American food products. Europeans also want to protect their small farms to maintain open space and rural employment. Such forces underlie much of the European anxiety about agricultural biotech - especially since it is seen as an "American invention."

Most of the industry leaders interviewed are quite enthusiastic about the benefits of biotechnology — especially in terms of increased food availability, enhanced nutrition, and environmental protection. Most feel that biotechnology has already provided benefits to consumers. Almost all recognise that foods developed through biotechnology have already been part of consumers' everyday diet. They clearly do not agree with most of the opponents' claims and tend to have almost no trust in such groups.

Their main concerns involve lack of consumer acceptance — not the safety of the foods. They express high levels of confidence in the science and the regulatory process. In fact, almost none feel that biotechnology should not be used because of uncertain, potential risks. Most food industry leaders do not feel it is necessary to have special labels on biotech-derived foods. They express concerns that such labels would be perceived as a warning by consumers. They also worried that the need to segregate commodities would pose financial and logistical burdens on everyone in the system - including consumers. Food industry leaders recognise a major need to educate the public about biotechnology. They look to third parties, such as university and government scientists to provide such leadership.

Research shows that consumers will accept biotech foods if they see a benefit to themselves or society and if the price is right. Their responses to foods developed through biotechnology are basically the same as for any other food - taste, nutrition, price, safety and convenience are the major factors that influence our decisions about which foods to eat. How seeds and food ingredients are developed will only be relevant for a relatively small group of concerned, consumers.

The potential for public concerns has led several food companies to change their products to avoid biotech-derived ingredients. For example, Gerber Foods received threats from Greenpeace because they had determined the company was using biotech-derived food ingredients (mainly soy). The company firmly believes that the biotech foods are safe to consume. Gerber agreed to drop some of its existing corn and soybean suppliers in favour of ones that can produce crops that are not genetically altered. It became an issue that is suddenly confronting other food companies. A private manufacturer in California, called Healthy Times Natural Food has switched from Canola oil (which sometimes is genetically modified) to safflower oil after facing questions from Greenpeace. The controversy is due in part to the fact that the organic industry is using public concern as a tool for marketing their products as free of biotech ingredients.

National and International Biotechnology Policy

National governments and international policy making bodies rely on food scientists and others to develop innovations that will create marketable food products and increase food supplies. Governments also rely on scientific research because they are responsible for setting health and safety standards regarding new developments. International organisations can suggest policy approaches and help develop international treaties that are ratified by national governments.

Economic success in the competitive international market demands that food production become more efficient and profitable. National governments and international organisations support food biotechnology as a means to avoid global food shortages. Many policy-making bodies are also trying to balance support of the food biotechnology industry with public calls for their regulation. Such regulations are necessary to protect public health and safety, to promote international trade, conserve natural resources, and account for ethical issues.

The majority of processed foods on the market contain soy or corn ingredients that come from GM plants. To date none have posed a food safety risk. The chief safety concerns are the potential to alter nutrient content or introduce allergens. Federal agencies involved in biotechnology regulation include the US Department of Agriculture (USDA) which evaluates agricultural production processes for all foods; the Food and Drug Administration (FDA), which evaluates whole non-animal foods (seafood), food ingredients, and food additives; and the Environmental Protection Agency (EPA), which evaluates plants with insecticidal properties.

Developers of GM plants and biotech-derived foods are required to consult with FDA prior to the commercialisation of the product. This consultation procedure entails a science-based safety assessment of the product that focuses on the protection of the consumer, developer, and the environment. Thus developers, have a strong incentive to cooperate fully with FDA and the other agencies prior to marketing their products.

Labelling Issues

Labelling food derived from GM plants and animals is an important, but complex issue. Some consumers and consumer groups believe they have a right to know whether biotechnology was used to produce food. Others believe labelling is not necessary if foods are essentially equivalent in composition. Food labels are regulated by FDA and, in some cases, by USDA. Regulatory agencies are concerned with ensuring that food labels are both true and not misleading. In accordance with their statutory mandate, the FDA has determined that a food product should be labelled as a product of biotechnology only if it has been changed in some significant way. The US Court of Appeals recently reaffirmed that the FDA's labelling policy is correct.

Based on the input from three public hearings, the FDA has recognised that the most effective way to allow for informed choice will be a system of voluntary labelling for foods not produced through biotechnology. If the demand for "GM-free" food is real, a market will become viable. In this case, meaningful choice can be provided to concerned consumers without imposing costs on or denying benefits to the majority of consumers, who support biotechnology. This may be the only viable and cost-effective option for the food industry (given the costs and difficulties associated with segregation and identity preservation.)

National surveys conducted with American consumers between 1997 and 2000 found about three-quarters of consumers supported this FDA labelling policy. There is evidence from focus groups that US consumers are already overwhelmed by the level of detail on food labels and mainly look for relevant information about nutrition. Recent focus groups have demonstrated that the wording on labels has a significant effect on consumer understanding and acceptance of biotechnology. Any label messages need to be simple and clear. However, labelling will be counterproductive without education.

It is important to understand when FDA will and won't require a product label. The same labelling laws that apply to all other foods and food ingredients apply to products of food biotechnology. Requirements for all food labels mandate proper identification of products and notice of health or safety concerns. Labelling would be required in the following instances - but not simply because the products were made using biotechnology.

- Potential food allergy is an example of a health or safety risk that would mandate a product label (e.g., if a gene has been transferred from a known allergen, such as a peanut).
- When substantial changes have been made to the expected nutritional composition of a food (e.g., vitamins have been added to a particular type of produce).
- If the common identity of a food product has been changed (e.g., if the composition of a fruit or vegetable becomes significantly different it would have to have a different name).

Labelling of biotech-derived foods will be required in some countries (such as Japan and the European Union) for cultural and religious reasons. These countries also believe labelling should be required simply because the consumers want to know more about the content of the food or how it was produced. Some have suggested that such labelling initiatives have been driven by economic and political interests rather than by scientific need. Consumers in these countries will ultimately end up paying even higher prices for their food as a result of such strict labelling policies.

The applications of biotechnology are so broad, and the advantages so compelling, that virtually every industry is using this technology. Developments are underway in areas as diverse as pharmaceuticals, diagnostics, textiles, aquaculture, forestry, chemicals, household products, environmental cleanup, food processing and forensics to name a few. Biotechnology is enabling these industries to make new or better products, often with greater speed, efficiency and flexibility. Biotechnology holds significant promise to the future but certain amount of risk is associated with any area. Biotechnology must continue to be carefully regulated so that the maximum benefits are received with the least risk.

Biotechnology is at a crossroads in terms of public acceptance. Many Americans have not yet formed a solid opinion on this complex issue. International developments over the next few years will certainly have a major influence on the long-term viability of biotechnology. The future of the world food supply depends upon how well scientists, government, and the food industry are able to communicate with consumers about the benefits and safety of the technology.

Several major initiatives are under way to strengthen the regulatory process and to communicate more effectively with consumers. Both the USDA and FDA have opened their regulatory systems to outside review and public comment. The biotechnology industry, university scientists and others are also conducting educational programmes. These should further strengthen consumer confidence. This partnership among the public and private sectors will support these emerging technologies that will prove vital to the US economy and the developing world in the new millennium. Even Europe will soon find the real benefits of biotechnology compelling.

Environmental biotechnology, more than just genetic engineering Biotechnology is the development of products or processes using plants, animals or microorganisms. Sometimes this involves 'genetic engineering' (altering the genetic material of living things so that they can produce new substances or perform new functions). This fact sheet describes some of the many clever non-genetic engineering biotechnologies that can be used to protect the environment better and conserve biodiversity.

Cleaner production - manufacturing with less pollution or less raw materials Environmental biotechnology offers ways to make industrial processes work more efficiently and create less pollution.

- Biotechnology may make available biological replacements for synthetic chemicals. For example, it is possible to make a variety of plastics from plant sugar rather than petrochemicals by using specially tailored yeasts and microorganisms. The advantage of these products is that they are biodegradable.
- It is expected that bioleaching will progressively displace chemicals used in recycling or bleaching paper. Many laundry detergents use enzymes (although in some cases these may be produced using gene technology) to replace phosphate detergents.
- Some heavy-duty stain removing bacteria have even been found in heavily-alkaline lakes where they have survived by learning to break down the toxins in their environment.
- Scientists are looking for microorganisms in hostile natural environments (for example very hot, cold or oily places) to use the talents they have developed for industrial uses.

Better Treatments for Solid Waste and Wastewater

- Most sewage treatment plants traditionally use a combination of chemical, physical and microbial treatment to break down waste. It is now possible to use modern biotechnology techniques to analyse the conditions needed to optimise the performance of the bugs and the systems, and to tailor the technology to different uses, from backyard septic tanks to large-scale intensive animal farm such as piggeries.

- The CSIRO has used enzymes to treat run-off from farms that contains pesticides, to avoid contaminating rivers and downstream farms.
- Sometimes the 'treatment' of waste produces valuable products, and modern high-tech composting factories can turn tons of organic garbage into precious soil in just days.

Bioremediation-Cleaning up Contamination

- Naturally occurring microorganisms, mainly bacteria and fungi, are being used to help clean up some of Australia's 60,000 sites contaminated by heavy metals, acids, petroleum derivatives, chlorinated solvents and explosives.
- Certain plants have also been found to absorb toxic metals such as mercury, lead and arsenic from polluted soils and water, and scientists are hopeful that they can be used to treat industrial waste.
- Oil spills, from small industrial puddles to massive ocean oil spills are often treated with oil-eating bacteria.

Tracking the Health of the Environment through Biomonitoring

Animals react to their environments, and can be used to gauge problems.

- The livers of sand flathead have proven to be useful indicators of municipal and industrial pollution from dioxins and other hazardous chemicals.
- Two frequently harvested species of marine mollusc, the Sydney rock oyster and the mussel, may be useful 'bioindicators' for the heavy metal toxins zinc and cadmium.

Biomass Energy

'Biomass' is plant and animal material that can be used as an energy source, from the traditional wood to waste material such as bagasse from sugarcane, to specially grown energy crops that can be converted to ethanol using modern biotechnology techniques and used with petrol in vehicles.

- Several Australian companies are planning to build biodiesel plants that convert vegetable oil or abattoir by-products into diesel fuel. Another company has a technology that converts green waste from rubbish tips into electricity.
- Even human waste can be converted into a combustible fuel.

Plants: Genetic engineering is not the only tool for creating novel plants with special traits. Conventional breeding is still used for developing useful plants, and which can be done with fewer risks.

- By cross breeding, scientists are developing wheat plants to increase their salt tolerance so they can grow in areas of greater salinity.
- Eucalypts are also being cross-bred to introduce natural genes from other species to create plants that tolerate salinity and dryland conditions.

Genetic Engineering for Environmental Solutions

Many of the biotechnology applications described above can be varied or refined using genetic engineering.

- CSIRO and Orica Australia Ltd. are using gene technology to develop enzyme products that detoxify pesticide residues, which in Australia would be of particular value to the cotton, horticultural and rice industries.
- In a variation on bioindicators, bacteria have been genetically modified as 'bioluminescors' that give off light in response to several chemical pollutants. These are currently being used to measure the presence of some hazardous chemicals in the environment.
- Other genetic sensors that can be used to detect various

chemical contaminants are also being trialled. Some can be used to track how pollutants are naturally degrading in ground water.

- Many genetically modified plants have environmental benefits such as reduced pesticide use.

Applications of Biotechnology for Functional Foods

A relatively recent concept in the US to describe the broad healthfulness of foods is the term "functional foods." These foods are defined as foods that provide health benefits beyond basic nutrition (International Food Information Council 2004). The Food and Nutrition Board of the National Academy of Sciences described a functional food as, "any modified food or food ingredient that may provide a health benefit beyond that of the traditional nutrients it contains" (Food and Nutrition Board 1994). The original concept of functional foods originated in Japan from its development of a special seal to denote Foods for Specified Health Use (FOSHU). More than 270 foods have FOSHU status in Japan.

Foods qualify as "functional foods" because they contain non-essential substances with potential health benefits. Examples of the diverse foods and their bioactive substances that are considered "functional foods" are: psyllium seeds (soluble fibre), soy foods (isoflavones), cranberry juice (proanthocyanidins), purple grape juice (resveratrol), tomatoes (lycopene), and green tea (catechins). The broad classification of functional foods carries some irony, as John Milner, Chief of the Nutrition Science Research Group at the National Cancer Institute noted, "It is unlikely that a non-functional food exists."

Bioactive components of functional foods may be increased or added to traditional foods through genetic engineering techniques. An example would be the high lycopene tomato, a genetically modified tomato with delayed ripening characteristics that is high in lycopene, which has potent anti-oxidant capabilities. This report focuses on biotechnology applications in functional and improved foods, using the National Academy of Sciences definition as a guideline.

Applications of Biotechnology in Food Crops

In 1990, the US Food and Drug Administration (FDA) approved the first genetically engineered food ingredient for human consumption, the enzyme chymosin, used in cheesemaking.

It is estimated that today 70 per cent or more of cheese made in the US uses genetically engineered chymosin. The first genetically engineered food, the FlavrSavr tomato, was approved for human consumption in the US in 1994.

Transgenic Acreage Expands Steadily

Seven million farmers in 18 countries now grow genetically engineered crops. Leading countries are the US, Argentina, Canada, Brazil, China, and South Africa. Cultivation of genetically engineered crops globally has expanded more than 10 per cent per year for the past seven years, according to the International Service for the Acquisition of Agri-biotech

Applications (ISAAA, James 2004). Such an expansion rate amounts to a 40-fold increase in the global area of transgenic crops from 1996 to 2003. Thus, in spite of continuing controversy, the technology continues to be adopted by farmers worldwide. ISAAA highlighted its key findings this way:

> In 2003, GM crops were grown in 18 countries with a combined population of 3.4 billion, living on six continents in the North and the South: Asia, Africa and Latin America, and North America, Europe and Oceania. The absolute growth in GM crop area between 2002 and 2003 was almost the same in developing countries (4.4 million hectares) and industrial countries (4.6 million hectares). The three most populous countries in Asia—China, India, and Indonesia, the three major economies of Latin America—Argentina, Brazil and Mexico, and the largest economy in Africa, South Africa, are all officially growing genetically engineered crops.

The leading genetically engineered crops globally and in

the US are soy, maize (corn), cotton, and canola. In the US, transgenic virus-resistant papaya and squash are also cultivated.

Agronomic Traits Prevail

Research in plant biotechnology has focused primarily on agronomic traits—characteristics that improve resistance to pests, reduce the need for pesticides, and increase the ability of the plant to survive adverse growing conditions such as drought, soil salinity, and cold.

Biotechnology traits developed and commercialised to date have largely focused on pest control (primarily Bt crops) or herbicide resistance. Many plant pests have proven either difficult or uneconomical to control with chemical treatment, traditional breeding, or other agricultural technologies and in these instances in particular, biotechnology has proven to be an effective agronomic tool. Herbicide resistance allows farmers to control weeds with chemicals that would otherwise damage the crop itself.

Varieties combining two different traits, such as herbicide tolerance and insect resistance, have been introduced in cotton and corn. The addition of new traits, such as resistance to rootworm in maize, and the combinations of traits with similar functions, such as two genes for resistance to lepidopteran pests in maize, are expected to increase. In its 2003 report, ISAAA suggested that five new Bt and novel gene products for insect resistance in maize could be introduced.

While the improvement of agronomic characteristics in major crops has been highly successful, few products genetically engineered to meet the specific needs of either food processors or consumers have yet been commercialised. Recently, however, a renewed emphasis on developing agricultural biotechnology applications more relevant to consumers has accompanied continuing efforts to develop crops with improved agronomic traits. Although genetically engineered crops with enhanced health, nutrition, functional, and consumer benefits have lagged behind agronomic applications, research on many such products is in the advanced stages of development. These applications could improve human and livestock nutrition and health, the nutritional quality of food animals for human consumption, and create ingredients with superior properties for food manufacturing and processing.

Food applications for Human Health

Quantity and Quality of Food Oils

Food oils have both nutritional and functional qualities. From a nutritional perspective, fats and oils contribute more energy (calories) than any other nutrient category, about nine calories per gram. This compares with about four calories per gram from carbohydrates and protein. At the same time, specific fatty acids that comprise most of what we call "fat" can affect a person's risk of developing certain chronic diseases such as heart disease.

Research over the past several decades has shown that some categories of fatty acids, such as saturated fatty acids, increase the risk of heart disease and other chronic diseases when consumed in excess. Fatty acids also influence how foods behave during manufacturing and processing. For example, saturated fatty acids add stability, texture, and flavour to foods, so they are not simple to replace.

To reduce the saturated fatty acid content of foods, plant breeders and food manufacturers increased their use of vegetable oils rich in polyunsaturated fatty acids and developed food oils low in saturated fatty acids. One example is canola oil with 6 per cent to 7 per cent total saturated fatty acids. To improve the stability of vegetable oils rich in polyunsaturated fatty acids, food manufacturers developed partially hydrogenated oils. The process of hydrogenation reduced the polyunsaturated fatty acid content and increased oil stability, but created *trans* fatty acids, which were subsequently associated with adverse health effects. As a result, hydrogenated fats, the main source of dietary *trans* fatty acids, are now being eliminated from foods. Food manufacturers are developing other ways to reduce undesirable saturated fat content while maintaining stability such as using short chain saturated fatty acids and monounsaturated fatty acids.

One functional food oil created with the tools of biotechnology has been commercialised. Calgene's high lauric acid canola, Laurical, containing 38 per cent lauric acid, is used in confectionary products, chocolate, and non-food items such as shampoo.

Conventional canola oil does not contain lauric acid. Laurical is a substitute for coconut and palm oils. FDA approved its use in foods in 1995 (FDA 1995).

The more double bonds there are, the more unsaturated the fatty acid is. Fatty acids with just one double bond are called "monou".

Strategic Aims of Altered Fatty Acid Profile

Improving the healthfulness and functionality of food oils can be accomplished in several ways. Where traditional plant breeding reaches its limits, biotechnology may be used to:

- Reduce saturated fatty acid content for "heart-healthy" oils.
- Increase saturated fatty acids for greater stability in processing and frying.
- Increase oleic acid in food oils for food manufacturing.
- Reduce alpha-linolenic acid for improved stability in food processing.
- Introduce various omega-3 polyunsaturated fatty acids including long-chain forms.
- Enhance the availability of novel fatty acids, e.g., gamma-linoleic acid.

Achievements in Altered Fatty Acid Profile

Reduced Saturated Fatty Acid Content: Genetically modified soybeans have been developed that contain about 11 per cent saturates compared with 14 per cent in conventional soybeans. In May 2003, scientists reported the development of transgenic mustard greens (*Brassica juncea*) containing 1 per cent to 2 per cent saturated fatty acids, a level significantly less than in the control plants (Yao *et al.* 2003). The transgenic plants also contained slightly higher amounts of oleic acid, a monounsaturated fatty acid, and higher levels of the polyunsaturates, linoleic and alpha-linolenic acids than the control plants. These results illustrate that alterations in one type of fatty acid may affect the levels of others, suggesting that combined strategies or genetic transformations may be necessary to achieve specific fatty acid profiles.

Palm oil low in saturated fatty acids is currently in development. This tropical oil contains about half saturated fatty acids (49.3 per cent), primarily palmitic acid (16:0, 43.5 per cent). However, with the recent success of biotechnology techniques in palm, transgenic palm oil enriched in oleic and stearic acids is under development (Parveez *et al.* 2000). Because of the long life cycle of palm and the time required to regenerate the plants in tissue culture, genetically engineered palm is not anticipated for another two decades (Parveez *et al.* 2000).

***Increased Saturated Fatty Acid Content*:** Because saturated fatty acids confer certain functional properties to food fats and oils and are more stable to heat and processing than unsaturated fatty acids, their use in cooking and baking is essential. To avoid the use of animal fats and hydrogenated vegetable oils with trans fatty acids, genetic engineering techniques have been used in canola and soy to develop oils with more short chain saturated fatty acids—12 to 18 carbons long—mainly lauric (12:0), myristic (14:0), palmitic (16:0), and stearic (18:0) acids. For example, Calgene's high lauric acid canola, Laurical™, containing 37 per cent lauric acid, was developed using the enzyme acyl-ACP thioesterase isolated from the California Bay Laurel (*Umbellularia californica*). Conventional canola oil contains no lauric acid, and only about 6 per cent short chain saturated fatty acids. This was the first transgenic oilseed crop produced commercially. High laurate canola is used in confectionary products, chocolate, and non-food items such as shampoo as a substitute for coconut and palm oils. FDA approved its use in 1995.

Enrichment of canola with even shorter chain saturated fatty acids, those with eight and ten carbons, has also been accomplished (Dehesh *et al.* 1996). Using a palmitoyl-acyl carrier protein thioesterase gene from a Mexican shrub, *Cupea hookeriana,* Dehesh and colleagues developed lines of canola with as much as 75 per cent caprylic (8:0) and capric acids (10:0). These fatty acids are absent in conventional canola oil. When consumed, these water-soluble fatty acids are mainly oxidised for energy.

Soybeans have been genetically modified to produce oil enriched in stearic acid (18:0), a saturated fatty acid that scientists believe does not raise serum cholesterol levels. The stearic acid-rich oil had 28 per cent stearic and 20 per cent oleic acids, with lower linoleic acid (18:2) than the conventional oil. Gene transfer technology also boosted the stearic acid content of canola (Hawkins and Kridl 1998).

Researchers at Calgene, Inc., Davis, CA, cloned three thioesterase genes from mangosteen, a tropical tree that stores up to 56 per cent of its seed oil as stearate. One of these genes led to the accumulation of up to 22 per cent stearate in transgenic canola seed oil, an increase of more than 1,100 per cent over conventional varieties (Hawkins and Kridl 1998).

Increased Oleic Acid Content: The most recent approach to develop more healthful food oils is increased oleic acid content. High oleic acid oils are lower in saturated and polyunsaturated fatty acids compared with conventional oil. Oleic acid, the predominant monounsaturated fatty acid in seed oils, is abundant in olive (72 per cent), avocado (65 per cent), and canola (56 per cent) oils, but not in others. Like saturated fatty acids, high oleic acid oils are useful in food processing and manufacturing for maintaining functionality and stability during baking and frying. Unlike saturated fatty acids, however, they do not raise blood cholesterol concentrations and are therefore, considered more healthful.

Biotechnology offers a means to increase the oleic acid content of vegetable oils, usually at the expense of polyunsaturated fatty acids, and sometimes, saturated fatty acids, depending on the particular transformations used. The concomitant reduction in polyunsaturated fatty acids has the added advantage of increasing the stability of the oil and ultimately the processed food. While traditional plant breeding allowed a modest increase in oleic acid, biotechnology has been necessary to achieve the high levels desired. For example, canola oil moderately high in oleic acid was developed using traditional plant breeding techniques.

With the application of biotechnology, oleic acid content increased to 75 per cent (Corbett 2002). Others have developed canola oil with more than 80 per cent oleic acid (Wong *et al.* 1991, Scarth and McVetty 1999).

More recently, Buhr and colleagues at the University of Nebraska used genetic engineering to increase oleic acid levels in soybeans by inhibiting the ability of the plant to convert oleic acid to polyunsaturated fatty acid (Buhr *et al.* 2002). When the conversion enzyme was inhibited, the level of oleic acid increased from 18 per cent in the wild-type seed to 57 per cent in the transgenic seed. When two gene transformations were applied, oleic acid content increased to 85 per cent, with saturated fatty acids reduced to 6 per cent.

Using a different approach, scientists at DuPont used the technique of cosuppression to reduce the production of polyunsaturated fatty acids in soybeans. Cosuppression occurs when the presence of a gene silences or turns off the expression of a related gene. Like Buhr and colleagues, these scientists were able to turn off the production of the enzyme that converts oleic acid to polyunsaturated fatty acids. The result was greatly increased production of oleic acid and reduced production of polyunsaturated fatty acids.

Gene silencing has also been used to produce high oleic and high stearic acid cottonseed oils (Liu *et al.* 2002). Cottonseed oil is high in palmitic acid, very high in linoleic acid, and free of alpha-linolenic acid. Conventional cottonseed oil has about 13 per cent oleic acid. When gene silencing was used to transform cotton, the resulting oil had 78 per cent oleic and only 4 per cent linoleic acids, respectively, with palmitic acid reduced from 26 per cent to 15 per cent. Cotton was also genetically modified to produce high stearic oil having 40 per cent stearic and 39 per cent linoleic acids, with 15 per cent palmitic acid. A combination was also developed to have 40 per cent stearic, 37 per cent oleic and only 6 per cent linoleic and 14 per cent palmitic acid. These examples illustrate the power and specificity of this technology to develop tailored seed oils.

Reduced Alpha-linolenic Acid: Several genetic transformations designed to increase oleic or stearic acid content do so at the expense of the polyunsaturated fatty acids alpha-linolenic and linoleic acids. These fatty acids have desirable nutritional characteristics, but their presence reduces the stability of oils for baking, processing, and frying and increases their susceptibility to oxidation or rancidity. Oils with appreciable amounts of alpha-linolenic acid such as canola and soybean, with about 10 per cent and 8 per cent alpha-linolenic acid, respectively, have been genetically modified to reduce this fatty acid. Such oils would be desirable for the commercial uses mentioned. Pioneer Hi-Bred, a DuPont company, developed low alphalinolenic acid soybean seeds through conventional breeding techniques with less than 3 per cent alpha-linolenic acid in its oil. Marketed under the brand TREUS the company claims that the oil eliminates the need for hydrogenation in food processing. A similar product from Monsanto, Vistive, offers a similar level of reduction in alpha-linolenic acid.

Omega-3 Fatty Acids: There is extensive interest in increasing Americans' consumption of omega-3 fatty acids, because they are associated with many health benefits, but are consumed only in small amounts. In 2002, the National Academy of Sciences Institute of Medicine recognised that omega-3 fatty acids are essential in the diet and established an estimated adequate intake for them (Institute of Medicine 2002). The main food sources of the long-chain omega-3 fatty acids are fish, especially fatty species such as salmon, rainbow trout, mackerel, herring, and sardines. Some plants—mainly canola, soybean, and flax oils—provide the 18-carbon omega-3 fatty acid, alpha-linolenic acid.

However, higher plants lack the enzymes to make 20- and 22-carbon polyunsaturated fatty acids needed by mammals. Humans can convert alpha-linolenic acid to the more biologically active longchain forms, but they do so very inefficiently. Thus, plant foods with alpha-linolenic acid may be insufficient to supply the need for long-chain omega-3 fatty acids, especially during pregnancy and lactation (Pawlosky *et al.* 2001). uinsaturated" and the amount in a food appears on the nutrition label. Olive oil and high oleic sunflower oil contain mainly monounsaturated fatty acids.

Other vegetable and fish oils are abundant in polyunsaturated fatty acids with two to six doubl.

Western diets contain predominately omega-6 polyunsaturated fatty acids found in soybean, corn, sunflower, canola, and cottonseed oils. It is now recognised that diets high in omega-6 fatty acids and low in omega-3 fatty acids may exacerbate several chronic diseases (Simopoulos *et al.* 2000). Because of the many health benefits associated with the regular consumption of omega-3 fatty acids, several health organisations, including the American Heart Association and the 2005 Dietary Guidelines for Americans, have called for increased consumption of these substances. One limitation to boosting consumption is that they occur naturally mainly in fatty fish and some seeds. Ironically, reducing the level of alpha-linolenic acid in soy and canola oils used in food processing, may actually reduce consumption of this fatty acid, although product developers are working to combine high omega-3 and low alpha-linolenic traits in one product.

Although aquaculture has increased the availability of some fish and shellfish species, increasing worldwide demand has put severe pressure on wild aquatic resources and limited seafood availability. Thus, it would be desirable to increase the availability of these fatty acids or their precursors in a variety of other foods, especially plants. Such foods would also be useful for animal and fish feed.

One strategy to increase the availability of long-chain omega-3 fatty acids is to develop oilseed crops such as canola and soybean that contain *stearidonic acid*. This omega-3 fatty acid occurs naturally in only a few plants such as black currant seed oil and echium. Stearidonic acid is the first product formed when alpha-linolenic acid is converted to eicosapentaenoic acid (EPA), a desirable long-chain omega-3 fatty acid. Usually, this first step limits the amount of EPA produced, but increasing the level of stearidonic acid helps overcome this limitation. Then the body's enzymes convert stearidonic acid to 20-carbon polyunsaturated fatty acids.

Dr. Virginia Ursin and colleagues at Calgene studied the metabolism of stearidonic acid in people (James *et al.* 2003). Her studies showed that when either stearidonic acid or EPA was consumed the amount of EPA in red blood cells increased significantly. This finding meant that the stearidonic acid was converted to EPA and appeared in red cells just as readily as the preformed EPA.

In contrast, when the study volunteers consumed alpha-linolenic acid, there was no change in their red cell EPA content. None of the fatty acids consumed had any effect on cell DHA levels, another long-chain omega-3 fatty acid associated with health benefits. Although the study used supplements, not stearidonic acid from transgenic plants, the findings suggest that plants with stearidonic acid would have potential to provide EPA.

Towards this end, scientists at Calgene, have successfully transformed canola so that it makes stearidonic acid. This genetic engineering feat required two genes from the fungus *Mortierella alpina* and one from canola for the three enzymes needed to produce sufficient stearidonic acid (Ursin 2003). The engineered plants accumulated up to 23 per cent stearidonic acid in the seed oil with a reduction in oleic acid content from about 60 per cent to about 22 per cent.

By breeding the transgenic lines with various lines of canola the investigators were able to develop a line of canola containing more than 55 per cent of alpha-linolenic acid and stearidonic acid. Total omega-6 fatty acids remained about 22 per cent, a level similar to conventional canola.

Calgene scientists have also developed soybean that contains stearidonic acid (Ursin, personal communication 2004).

The implications of Calgene's work with stearidonic acid are substantial. This is the first demonstration of the incorporation into edible plants of a biologically potent source of long-chain omega-3 fatty acids. This work marks an important advance in the development of plant-based sources of long-chain omega-3 fatty acids that could be consumed directly or incorporated into food products. However, because stearidonic acid contains four double bonds, it is vulnerable to oxidation and would require anti-oxidant protection. One can imagine that transgenic canola and soybean could be developed using additional traits to boost anti-oxidant protection, possibly from vitamin E.

In May 2004, a landmark paper announced the production of long-chain polyunsaturated fatty acids—both omega-6 and omega-3 types—in *Arabidopsis thaliana*, a type of cress widely used as a model plant in biotechnology research. Dr. Baoxiu Qi and colleagues at the University of Bristol, United Kingdom, transferred to *Arabidopsis thaliana* three genes encoding for different enzymes in the metabolic pathway from linoleic and alphalinolenic acids to arachidonic and eicosapentaenoic acids, respectively (Qi *et al.* 2004). The additional genes were necessary to provide the enzymes to make these long-chain fatty acids. Yields of EPA (13 per cent) and arachidonic acid (29 per cent) in leaves were significantly higher than in conventional cress, which usually does not produce these fatty acids, and accounted for 43 per cent of the total 20-carbon polyunsaturated fatty acids. In addition to the production of EPA and arachidonic acid, the concentration of alpha-linolenic acid was reduced from 48 per cent to 14 per cent. This achievement was also remarkable because it used a pathway seldom found in plants.

This work is important in several regards. One is that it demonstrates the feasibility of developing plants capable of synthesising long-chain polyunsaturated fatty acids. Another is the relatively high efficiency of conversion of the precursor fatty acids to the long-chain forms. A third advantage is the improved balance of omega-6 and omega-3 fatty acids, with significant reduction in the amounts of the 18-carbon precursors linoleic and alphalinolenic acid compared with conventional plants. Yet another is the demonstration that plants can be engineered not only with respect to the outcome of final products, but also the pathways for achieving the desired ends. A likely next step will be to apply this technology to seed oil crops such as canola and soybean to see if the long-chain polyunsaturated fatty acids will accumulate in the seed.

What distinguishes fats from one another is their fatty acids. Although production of EPA in plants represents an enormous scientific achievement, the question of making Docahexenoic Acid (DHA), a 22-carbon polyunsaturated omega-3 fatty acid important in retina and brain function and other body systems remained unsolved.

In mammals, the conversion of EPA to DHA is inefficient and requires several steps. It is possible, in theory, to perform this conversion in a direct manner, but the enzymes to do so are not present in mammals. Several research groups have examined many algae and identified the specific enzymes for this conversion (Sayanova and Napier 2004, Meyer *et al.* 2004).

Once the genes for these enzymes were identified and cloned they could be incorporated into model organisms to see whether DHA would be produced. In late 2004, Amine Abbadi at the University of Hamburg, Germany, working with colleagues in the UK and the US, reported the successful transformation of yeast that yielded small amounts of DHA (Abbadi *et al.* 2004). This accomplishment required four gene transformations. The team then went on to develop transgenic flax, a plant with abundant alpha-linolenic acid for conversion to long-chain fatty acids.

People have been warned for years to limit their intake of saturated fat, the kind rich in saturated fatty acids. These warnings relate to the ability of most saturated fat. Several steps remain before long-chain polyunsaturated fatty acids will be available in commercial crops. However, the demonstration that plants can be modified to make these important nutrients means that many of the scientific hurdles have been conquered. This work gives a large boost to the potential for plants to be an important dietary source of these fatty acids.

Gamma-linolenic Acid: This fatty acid is the first step in the conversion of linoleic acid to arachidonic acid in the omega-6 fatty acid pathway. When consumed in evening primrose or borage oils, it is poorly converted to arachidonic acid. For that and other reasons, it may have potential benefit in cardiovascular disease (Fan and Chapkin 1998). Gamma-linolenic acid has been associated with improved skin conditions in human subjects, improved liver function in patients with liver cancer, and with anti-cancer effects in cell culture studies. It was also shown to enhance the effectiveness of tamoxifen, an anti-estrogenic medication used to prevent the recurrence of breast cancer. It is believed to suppress the production of estrogen receptors in cells.

Butter, cheese and other dairy foods, and meats are rich in saturated fatty acids. Gamma-linolenic acid is naturally present in appreciable amounts in few plants, notably borage

(*Borago officinalis*), evening primrose (*Oenothera biennis*), black currant oil (*Ribes nigrum*) and echium (*Echium plantagineum*). The ability to increase the production of gamma-linolenic acid in tobacco plants by transferring the gene for the delta-6 desaturase enzyme from various sources was first shown in 1996 by Reddy and Thomas at Texas A&M University, and by others in 1997 (Reddy and Thomas 1996, Sayanova *et al.* 1997). A recent study reported that gamma-linolenic acid content in transgenic canola ranged from 22 per cent to 45 per cent (Wainright *et al.* 2003). Evening primrose has also been genetically modified for enhanced gamma-linolenic acid content (Wainright *et al.* 2003). Arcadia Biosciences, Davis, CA, has also reported transgenic safflower plants with 65 per cent gamma-linolenic acid in the oil.

Ursin's study of transgenic canola enriched in stearidonic acid also reported that a cross between the transgenic line of canola producing stearidonic acid and a canola line high in gamma-linolenic acid yielded a canola containing about 11 per cent gammalinolenic acid and about 14 per cent stearidonic acid (Ursin 2003). This example illustrates the variety of fatty acids that can be developed in seed oils using a combination of genetic engineering and traditional plant breeding techniques.

Quantity and Quality of Plant Protein

Efforts to improve the protein content and quality of staple foods have been underway for decades. The main focus is crops grown in developing countries, where nutrient shortfalls are widespread and dietary diversity limited. Foods such as potato and cassava, staple foods in several parts of South America and Africa, have less than one per cent protein.

Efforts to improve protein quality strive to increase the amount of limiting essential amino acids provided by the protein in the food. The amino acids most often present in inadequate amounts are lysine, tryptophan, and methionine. Improvements in protein quality benefit both human and animal nutrition and increase the feed efficiency of crops fed to food animals. For example, corn is widely fed to cattle but it is limiting in lysine and methionine.

Corn with higher levels of these amino acids would significantly improve feed efficiency and lower input costs to farmers. Improved corn varieties consumed by humans would also have nutritional benefits.

There are various ways of improving protein quantity and quality. One is to increase the total amount of protein produced by selecting germplasm with an altered balance of seed proteins. This may be done by traditional cross breeding or genetic engineering. Another approach is to introduce genes from other sources for proteins that have a favourable balance of essential amino acids. An example is the introduction into potato of a gene for seed albumin protein from amaranth. A third approach seeks to increase the production of specific amino acids such as lysine. This approach was used in the development of Quality Protein Maize.

William Folk and his team at the University of Missouri, Columbia, MO, pioneered another approach to improve seed protein quality. Their strategy was to substitute more desirable and scarce amino acids for more abundant ones in certain seed proteins (Chen *et al.* 1998, Wu *et al.* 2003). They applied this concept to rice by increasing the production of lysine, an essential amino acid, at the expense of the non-essential amino acids, glutamine, asparagine and glutamic acid.

Cassava: A staple food for some 500 million people in tropical and sub-tropical parts of the world, cassava (*Manihot esculenta* Crantz), also known as yucca or manioc, thrives in marginal lands having little rain and nutrient-poor soils. It is widely consumed in Africa, and parts of Asia and South America. Cassava root has less than 1 per cent protein and poor nutritional value. However, the leaves are also consumed and these are a good source of beta-carotene, the precursor of vitamin A.

In 2003, Zhang and colleagues reported using a synthetic gene to increase the protein content in cassava (Zhang *et al.* 2003). The gene is for a storage protein rich in nutritionally essential amino acids. When the gene was expressed in cassava, transformed plants expressed the gene in roots and leaves, both of which are consumed in human diets. The experiment demonstrated the feasibility of increasing the quantity and quality of protein in cassava.

Cassava also contains cyanogenic glucosides that can produce chronic toxicity if not eliminated or reduced by grating, sun-drying, or fermenting. Efforts to develop cassava varieties low in these toxicants is a high research priority.

Corn: Corn (*Zea mays*) is the predominant staple food in much of Latin America and Africa. Although some varieties may contain appreciable quantities of protein, its quality is poor because of low lysine and tryptophan content. In 1964, it was discovered that corn bearing a gene known as opaque-2 contained increased concentrations of lysine and tryptophan and had significantly improved nutritional quality (Food and Agriculture Organisation 1992). However, opaque-2 corn proved to have low yields, increased susceptibility to diseases and pests, and inferior functional characteristics.

At the International Maize and Wheat Improvement Centre (CIMMYT) in Mexico, work with the opaque-2 gene continued using both traditional breeding and molecular methods.

After at least 12 years' work, CIMMYT researchers succeeded in developing hardy corn varieties that contained twice the lysine and tryptophan content as traditional varieties, but were disease-resistant and high-yielding. Scientists Surinder K. Vasal and Evangelina Villegas of CIMMYT were awarded the World Food Prize in 2000 for their work developing 'Quality Protein Maize'. Quality Protein Maize varieties have been adapted to and released in over 40 countries in Latin America, Africa, and Asia.

Recent researchers at CIMMYT reported the development

of transgenic corn with multiple copies of the gene from amaranth (*Amaranthus hypochondriacus*) that encodes for the seed storage protein amarantin (Rascon-Cruz *et al.* 2004). Total protein in the transgenic corn was increased by 32 per cent and some essential amino acids were elevated 8 per cent to 44 per cent.

In 2004, a team of researchers at the University of California, Riverside, reported that transgenic corn with increased production of the plant regulating hormone, cytokinin, had nearly twice the content of protein and oil as conventional corn (Young *et al.* 2004). This development resulted from an unusual change in the way the plant developed. Normally, corn ears develop flowers in pairs, one of which usually dies. Under the influence of the additional cytokinin, both flowers developed but yielded only a single kernel. These kernels contained more protein and oil than conventional corn.

Pursuing a different strategy to improve protein quality, researchers at Monsanto, St. Louis, MO, used genetic engineering techniques to reduce the amount of zein storage proteins.

These storage proteins constitute over half the protein in corn and are deficient in lysine and tryptophan. Increased production of other proteins in the corn led to higher levels of lysine, tryptophan, and methionine (Huang *et al.* 2004). The agronomic and nutritional properties of these lines are currently being evaluated.

Researchers at the Max Planck Institute, Germany, have focused on methionine, another limiting amino acid. They elucidated several key steps in methionine metabolism in plants. This work, currently in the preliminary stage, could pave the way for using genetic engineering techniques to improve the methionine content of plants (Hesse and Hoefgen 2003).

Potato: Potato (*Solanum tuberosum*) is a dietary staple throughout parts of Asia, Africa, and South America. Typically, potatoes contain about 2 per cent protein and 0.1 per cent fat.

It was reported in 2000 that, as in cassava, transfer of the gene for seed albumin protein from *Amaranthus hypochondriacus* to potato resulted in a "striking" increase in protein content of the transgenic potatoes (Chakraborty *et al.* 2000).

In 2004, researchers at the National Centre for Plant Genome Research, India, reported the development of a nutritionally improved potato line with 25 per cent higher yields of tubers and 35 - 45 per cent greater protein content (ISAAA 2004). Dubbed the "protato," the protein-rich potato had significant increases in lysine and methionine, which enhance the quality of the additional protein (Council for Biotechnology Information 2004). In February 2004, this potato was reported "approaching release" to farming communities.

It should be noted that while potatoes are known for their high starch content, it has been possible to genetically engineer potatoes that contain fat (triglycerides). In July 2004, Klaus and colleagues at the Max Planck Institute of Molecular Plant Physiology demonstrated increased fatty acid synthesis in potatoes (Klaus *et al.* 2004).

Rice: Almost half the world's population eats rice (*Oryza sativa* L.), at least once a day (IRRI undated). Rice is the staple food among the world's poor, especially in Asia and parts of Africa and South America. It is the primary source of energy and nutrition for millions.

Thus, improving the nutritional quality of rice could potentially improve the nutritional status of nearly half the world's population, particularly its children. Commodity rice contains about 7 per cent protein, but some varieties, notably black rice, contain as much as 8.5 per cent (Food and Agriculture Organisation 2004). The most limiting amino acid in rice is lysine.

Efforts to increase the nutritional value of rice target protein content and quality along with key nutrients often deficient in rice-eating populations, such as vitamin A and iron.

The International Rice Research Institute (IRRI), Philippines, is a primary centre for rice research and development of improved varieties.

In 1999, Dr. Momma and colleagues at Kyoto University, Japan, reported a genetically engineered rice having about 20 per cent greater protein content compared with control rice (Momma *et al.* 1999). Transgenic plants containing a soybean gene for the protein glycinin contained 8.0 per cent protein and an improved essential amino acid profile compared with 6.5 per cent protein in the control rice.

As mentioned briefly above, Dr. William Folk and his team genetically modified rice to increase its content of the amino acid lysine (Wu *et al.* 2003). They did so by modifying the process of protein synthesis, rather than by gene transfer or the expression of new proteins. They achieved an overall 6 per cent increase in lysine content in the grain (Chen *et al.* 1998). Although lysine content remained below optimum levels, the scientists suggested that additional transformations and modifications could further boost lysine levels.

Perhaps the most famous genetic transformations in rice are those in "Golden Rice" involving the vitamin A precursor, beta-carotene, and iron. The lead scientist in the golden rice project, Dr. Ingo Potrykus, now retired from the Swiss Federal Institute of Technology, was also involved in applying biotechnology for the improvement of rice protein. Although details are sparse, Potrykus described the work of Dr. Jesse Jaynes, who synthesised a synthetic gene coding for an ideal high-quality storage protein with a balanced mixture of amino acids. The gene, named Asp-1, was transferred to rice with the appropriate genetic instructions for its production in the endosperm or starchy part of the rice grain.

The transgenic rice plants accumulated the Asp-1 protein in their endosperm in a range of concentrations and provided essential amino acids but data are not yet available on the concentrations achieved or their nutritional relevance. Precedent for the expression of a synthetic gene in rice grown in cell culture suggests that Jaynes' approach is viable (Huang *et al.* 2002).

Starch

Starch from cereals, grains, and tubers contribute a substantial share of dietary calories and in many poor countries, provide the majority of food energy. Starch is also important for feed and industrial purposes. Its use as paste goes back at least 4000 years BCE to the Egyptians who cemented strips of papyrus stems together with starch paste for writing paper.

Besides providing energy, starch confers functional characteristics to foods: texture, viscosity, solubility, gelatinisation, gel stability, clarity, etc. These characteristics depend on the proportion of amylose and amylopectin, the main components of starch. Amylose and amylopectin differ from each other in chain length, branching, and degree of polymerisation. Amylose is linear and amylopectin is highly branched. How a particular starch will be used in foods, determines what ratio of amylose to amylopectin is most suitable. High amylose starches include high amylose corn (70 per cent), corn (28 per cent), wheat (26 per cent) and sago (26 per cent). In contrast, waxy rice and waxy sorghum contain no amylose. Members of the potato family—potato, sweet potato, cassava—have 17 per cent to 20 per cent amylose.

Many lines of corn have been developed with different characteristics derived from modified starch ratios and increased amylose content. Transgenic high amylose potatoes developed by inhibiting two branching enzymes were reported to yield more tubers and have lower starch content, smaller granules, and increased reducing sugars (Hofvander *et al.* 2004).

Biotechnology has also been directed to increasing starch content (Geigenberger *et al.* 2001, Regierer *et al.* 2002). Potatoes were genetically altered to increase the activity of adenylate kinase, an enzyme involved in the plant's energy metabolism and starch production.

The resulting transgenic potatoes had substantially

increased adenylates and a 60 per cent increase in starch compared with wild-type plants (Regierer *et al.* 2002). Unexpectedly, the concentrations of several amino acids were increased 2- to 4-fold, and tuber yield increased.

Considerable publicity was given to potatoes engineered by Monsanto in the early 1990s to have increased starch content. These were touted as more desirable for French fries because they would absorb less fat during frying. They are an example of the type of starch modification that may have secondary health benefits as a consequence of how they are used.

Fructan

Fructans are polymers (repeating units) of the sugar fructose. They serve in food products as a low-calorie sweetener, source of dietary fibre, and bulking agent. They may also stimulate the growth of desirable colonic bacteria, such as bifida. Fructans have environmentally friendly non-food applications in the manufacture of biodegradable plastics, cosmetics, and detergents. Fructans are naturally occurring in Jerusalem artichokes (sunchokes) and chicory, but agronomic shortcomings in growing these crops have limited their use.

Inulin, a fructan found in Jerusalem artichokes, was successfully synthesised in potatoes following the transfer of two genes from globe artichokes (*Cynara scolymus*) (Hellwege *et al.* 2000). The full spectrum of inulin molecules present in artichokes was expressed in the transgenic potatoes. Inulin comprised 5 per cent of the dry weight of the transgenic tubers and did not influence sucrose concentration. However, starch content was reduced.

In a programme called the Agriculture and Fisheries Programme, or FAIR, the European Commission funded multidisciplinary research programmes in agriculture and fisheries, including a project on fructans for food and non-food uses. Research to date includes the isolation of several genes for fructosyl transferase enzymes involved in the production of fructans. The feasibility of using these enzymes has been demonstrated in model plants and target crops such as sugar beet (Anonymous 2000). In addition, it was reported in 2004 that genes encoding for fructosyl transferases in onion were isolated and transferred to sugar beet, a plant that does not normally synthesise fructans (Weyens *et al.* 2004). Following the transfer of the genes, onion-type fructans were produced from sucrose without loss in storage carbohydrate.

Increased Vitamin Content

Beta-carotene and other Carotenoids: Beta-carotene belongs to the family of carotenoids and is abundant in plants of orange colour. It is the precursor of vitamin A and can be converted to the active vitamin during digestion. Other carotenoids do not have potential vitamin A activity. Humans cannot synthesise carotenoids and therefore, depend on foods to supply them. Many staple foods, particularly rice, contain no beta-carotene or its precursor carotenoids. Diets lacking other food sources of vitamin A or beta-carotene are associated with vitamin A deficiency which can result in blindness, severe infections, and sometimes death. According to the World Health Organisation, vitamin A deficiency is the leading cause of preventable blindness worldwide. The deficiency affects some 134 million people, particularly children, in 118 countries. Overcoming this nutrient deficiency is an urgent global health challenge.

The development of "Golden Rice," so named because of its yellow colour conferred by the presence of beta-carotene, was a landmark achievement in the application of biotechnology to nutrition and public health. Peter Burkhardt, working with Ingo Potrykus and colleagues in Switzerland, was the first to show that transgenic rice, carrying a gene from daffodil, could express phytoene, a key intermediate in the synthesis of beta-carotene (Burkhardt *et al.* 1997). Subsequently, the Potrykus group reported the application of three transgenes in the development of rice expressing the entire pathway for the production of beta-carotene (Ye *et al.* 2000, Beyer *et al.* 2002). Additional work with Golden Rice included the insertion of a gene to increase the iron content (Potrykus 2003). IRRI is currently cross-breeding the nutrient-enhanced transgenic rice with local rice varieties from Asia and Africa, and fieldtesting the new lines for nutritional value and agronomic performance. Varieties of Golden Rice are not expected to be ready for farmers for several more years.

The development of transgenic plants able to produce a variety of carotenoids is an active area of research. It is clear that production of phytoene, the first product in the pathway for carotenoid synthesis, is the rate-limiting step in generating carotenoids (Cunningham 2002). Using gene transfer technology to increase the expression of phytoene synthase, the enzyme that makes phytoene, increases the synthesis of carotenoids substantially. For example, Shewmaker and colleagues (1999) from Monsanto reported an increase up to 50-fold in carotenoids, mainly alpha and beta-carotene, in canola (*Brassica napa*). However, vitamin E levels decreased significantly, oleic acid content increased, and linoleic and alpha-linolenic acids were reduced compared with non-transgenic seeds. These other changes would have to be modified or evaluated to determine whether they might have meaningful nutrition implications.

In a separate study on canola, the Monsanto group transferred to canola three genes from bacteria that affect the phytoene synthesis pathway. When they included a triple construct— genes for three different enzymes, phytoene synthase, phytoene desaturase and lycopene cyclase—the resulting transgenic canola seeds maintained the same amount of total carotenoids, but increased the ratio of beta to alpha-carotene from 2:1 to 3:1 (Ravanello *et al.* 2003).

Stalberg and colleagues in Sweden also studied the effects of phytoene synthase on carotenoid synthesis in transgenic *Arabidopsis thaliana*. They examined three ketocarotenoids; transformed seeds had a 4.6-fold increase in total pigment and a 13-fold increase in these three carotenoids (Stalberg *et al.* 2003). They also reported a 43-fold average increase in beta-carotene (Lindgren *et al.* 2003). Lutein, another nutritionally important carotenoid, was significantly increased, but zeaxanthin was only increased by a factor of 1.1. They also observed substantial levels of lycopene and alpha-carotene in the seeds, whereas only trace amounts were found in the control plants. However, germination was delayed in proportion to the increased levels of carotenoids.

Others have examined the effect of transgenes affecting phytoene metabolism on carotenoid synthesis. Dr. Peter Bramley's group at the University of London, United Kingdom, transformed tomatoes using a bacterial gene encoding for an enzyme that converts phytoene to lycopene, the precursor of beta-carotene. Tomatoes carrying the bacterial gene had about a 3-fold increase in beta-carotene content, but total carotenoids were not increased (Romer *et al.* 2000). The altered carotene content did not affect plant growth and development.

Lutein and zeaxanthin are nutritionally important carotenoids for protection of the retina and reduced risk of age-related macular degeneration (Krinsky *et al.* 2003, Gale *et al.* 2003).

Lutein is found in dark green leafy vegetables such as spinach and collards, and zeaxanthin occurs in yellow foods such as mangoes, corn, and peaches. The latter is not particularly abundant in Western diets. Romer and colleagues at Universitat Konstanz, Germany, were able to use biotechnology to block the conversion of zeaxanthin to another carotenoid and thereby increase its content in potatoes (Romer *et al.* 2002). With this approach they obtained increased levels of zeaxanthin in potatoes ranging from 4- to 130-fold. Total carotenoids were increased by 5.7-fold, but in some, lutein content was decreased. Alphatocopherol (vitamin E) was increased 2- to 3-fold in the transgenic potatoes. Fine-tuning these alterations has the potential to significantly enhance the nutritional value of potatoes. In another study, Bramley's group transferred the gene that increases carotenoid synthesis from a bacterium to tomatoes and measured total and specific carotenoids in the transgenic fruits (Fraser *et al.* 2002). Total carotenoids were 2- to 4-fold higher in transgenic fruits than in non-transformed plants, with increases in phytoene, lycopene, beta-carotene, and lutein of ranging from 1.8- to 2.4-fold.

Tomatoes with delayed ripening were produced as a result of inserting a gene encoding for S-adenosylmethionine decarboxylase, an enzyme involved in the ripening process (Mehta *et al.* 2002). An additional consequence of this transgenic modification was a severalfold increase in lycopene content. Lycopene is normally converted to beta-carotene, but tomatoes with increased lycopene content may have enhanced nutritional value. Lycopene consumption has been linked to reduced risk and spread of prostate cancer, though definitive data are lacking (Etminan 2004, Kristal 2004).

The carotenoid astaxanthin is synthesised by algae and plants and is responsible for the pink colour in shrimp and salmon. Humans absorb astaxanthin poorly, but absorption is increased in the presence of fat (Mercke Odeberg *et al.* 2003). Astaxanthin is of interest because of its strong anti-oxidant properties *in vitro*. It is less certain whether it is an anti-oxidant in human health. Astaxanthin is used commercially in feed for cultured salmon and trout.

Production of astaxanthin in flowers and fruits has also been accomplished with the techniques of biotechnology. Using a gene from the alga *Haematococcus pluvialis*, researchers at the Hebrew University of Jerusalem, Israel, transferred the gene into tobacco (*Nicotiana tabacum*). Transgenic tobacco plants produced astaxanthin and changed colour (Mann *et al.* 2000). This ability to manipulate pigmentation in fruits and flowers may have commercial potential and possible implications for increasing the availability of carotenoids for human health.

Vitamin E

There is strong interest in vitamin E because of studies linking it to decreased occurrence of several degenerative diseases and cancers, although efficacy remains unproven and data are inconsistent (71, 72). Some recent trials with vitamin E supplementation reported no protection against cardiovascular disease or cancer and some chance of increased risk of heart failure (Eidelman *et al.* 2004, Lonn *et al.* 2005, Miller *et al.* 2005). Also, because vitamin E is an anti-oxidant, it is useful in foods and oils to provide oxidative stability. Vitamin E is found mainly in vegetable oils, wheat germ, and a few other foods not widely consumed.

The most potent form of the vitamin is alpha-tocopherol, but the less potent gamma, beta, and delta forms are more widespread in plants. Efforts to increase the content of vitamin E in food plants, particularly cereals and grains, which may have low amounts, have sought to increase the amount of precursor substances by overexpressing the genes for various enzymes involved in the biosynthetic pathway.

Vitamin E biosynthesis involves complicated pathways so that multiple genetic manipulations are required. In September 2003, Dr. Edgar Cahoon of the US Department of Agriculture (USDA) and co-researchers at the Donald Danforth Plant Science Centre, St. Louis, MO, announced the development of transgenic corn with increased levels of vitamin E. Insertion of a gene from barley into corn increased the conversion of vitamin E precursors to vitamin E itself (Cahoon *et al.* 2003). The content of vitamin E and tocotrienol, a closely related substance, was increased up to 6-fold.

However, much of the anti-oxidant produced was tocotrienols rather than vitamin E (Aijawi and Shintani 2004). Tocotrienols, although potent anti-oxidants *in vitro*, are poorly absorbed in humans; however, they may have cholesterol lowering properties (Theriault *et al.* 1999). Besides enhancing the potential therapeutic and nutritional value of corn, this alteration may increase its oxidative stability after harvest.

In soybeans, Van Eenennaam and colleagues developed transgenic plants that were able to increase the conversion of the weaker forms of tocopherol typically present in soybeans to the more potent alpha-tocopherol (vitamin E) form. The result was a 5-fold increase in vitamin E activity (Van Eenennaam *et al.* 2003). This work paves the way for the development of vitamin E-rich oils and plants with potential health benefits.

Vitamin C

Vitamin C, or ascorbic acid, is abundant in citrus and other fruits such as strawberry and kiwi, but is very low in cereals and grains. Moreover, it cannot be synthesised by humans nor stored to any appreciable extent. Thus, we depend on regular dietary consumption to meet our vitamin C needs. In areas of the world where foods containing vitamin C are not widely consumed, strategies to increase the vitamin C in cereals and grains hold considerable potential to improve health.

In 2003, Gallie and colleagues at the University of California, Riverside, announced the successful transformation of corn and tobacco that resulted in a 20 per cent increase in vitamin C (Chen *et al.* 2003). They accomplished this increase by transferring a gene from wheat for an enzyme that recycles vitamin C and prevents its breakdown. Increased expression of this enzyme in transgenic corn and transgenic tobacco resulted in a 2- to 4-fold increase in vitamin C content in the kernel. Application of this approach to other food plants remains to be developed and evaluated.

The ability to increase the vitamin C content of strawberries, a good source of the vitamin, was reported in 2003 by a research team at the University of Malaga, Spain (Agius *et al.* 2003). Using a gene for the enzyme D-galacturonate reductase transferred to strawberry (*Fragaria spp*) this group showed that vitamin C content in the modified strawberry fruit increased with the expression of the transgene. This study demonstrated the feasibility of using this enzyme to raise the level of vitamin C in plants.

Folic Acid

Inadequate intake of the vitamin folic acid, one of a family of folates, is associated with megaloblastic anaemia, birth defects, impaired cognitive development, and some chronic diseases. Folates are available in small amounts in a variety of fruits and vegetables, but intakes tend to be low. For this reason, it would be desirable to increase the concentration of folates in dietary staples and foods widely consumed.

In the U.S. several foods are fortified with folic acid. In 2004, Hossain and colleagues at Tufts University, Medford, MA, and the Donald Danforth Plant Science Centre reported a 2- to 4-fold increase in folates and pterins, the precursors of folates, in transgenic *Arabidopsis thaliana* modified by the incorporation of the transgene for the first step in the synthesis of folic acid (Hossain *et al.* 2004). Other investigators have developed transgenic tomatoes, also enriched in the same gene, that had twice the amount of folate as control fruit (Diaz de la Garza *et al.* 2004). This group was able to boost folate content 10-fold by including a second gene transformation to increase the content of another substance, para-aminobenzoate, needed for folate synthesis. These studies provide good evidence of the potential to increase the availability of this vitamin in widely consumed foods.

Antioxidants

Vitamins C and E function as anti-oxidants in the body. However, many other substances widely distributed in foods in small amounts also provide protection against potentially damaging breakdown products arising from oxidation during normal metabolism.

Oxidation breakdown products, such as reactive oxygen species, oxidised lipids, and free radicals have been associated with chronic diseases, so there has been considerable interest in the availability of anti-oxidants. Caution should be sounded, however, because in small amounts many of these substances are protective; in high doses, they can act as prooxidants and may be harmful. Several examples of the applications of biotechnology for enhanced anti-oxidant capacity in foods are described below.

Phenolic compounds are the most widespread anti-oxidants in foods. They include such substances as flavanols, tocopherols, quercetin, resveratrol, and many others. They have become familiar to consumers because they are touted in foods as diverse as berries, wine, tea, olive oil, and many others. Potatoes are a source of anti-oxidant flavanoids and vitamin C. To enhance the anti-oxidant content of potatoes, Lukaszewicz and colleagues conducted a series of transformations using one or multiple genes encoding enzymes in the bioflavanoid synthesis pathway (Lukaszewicz *et al.* 2004). Transgenic plants exhibited significantly increased levels of phenolic acids and anthocyanins, plus improved anti-oxidant capacity. However, starch and glucose levels were decreased. These findings point to complex relationships between anti-oxidant content and other compounds, but indicate that anti-oxidant levels in potatoes can be altered using biotechnology.

Another phenolic anti-oxidant, chlorogenic acid, accumulates in some crops and is found in apples, green coffee beans, tomatoes, and tea. It is synthesised by the enzyme hydroxycinnamoyl transferase in solanaceous plants (e.g., potato, tomato, eggplant).

In 2004, it was reported that transgenic tomatoes carrying the gene for this enzyme accumulated higher levels of chlorogenic acid with no side effects on levels of other phenolics (Niggeweg *et al.* 2004). The transgenic tomatoes also showed improved anti-oxidant capacity, suggesting that such enhanced tomatoes might provide additional anti-oxidants.

Yet another transformation in tomatoes was recently reported to result in the synthesis of resveratrol, an anti-oxidant not normally found in tomatoes. Resveratrol is usually associated with grapes and wines where it is abundant. In this study, tomatoes incorporating the gene for stilbene synthase, an enzyme in the pathway for resveratrol synthesis, had a resveratrol content of 53 mg/g fresh tomato upon ripening (Giovinazzo *et al.* 2005). The contents of two other anti-oxidants, vitamin C and glutathione, were also increased.

Mineral Content and BioavaIlabiliIty

Improving human nutrition by increasing the availability of trace minerals in crops is potentially highly efficient and effective. This strategy may reach more people in developing countries than fortification of foods, because many subsistence farmers grow their own food and are outside the market system. If they have access to and consume improved crop varieties, they will not only improve their nutrient intake, they may improve their crop yields and consequently their economic wellbeing. This is because trace minerals are essential to the plant's ability to resist disease and other environmental stresses (Bouis 2002). Further, plants with improved ability to take up minerals from the soil will not deplete nutrient-poor soils. Such plants are able to unbind minerals in the soil and make them available to the plant, thus making use of an abundant resource in the soil that is otherwise unavailable. Mineral-efficient plants are also more drought resistant and require fewer chemical inputs (Bouis 2002).

Iron

Iron deficiency anaemia is one of the most widespread nutritional deficiencies in the world.

The United Nations estimates that over three billion people in developing countries are iron-deficient (Administrative Committee 2000). The problem for women and children is more severe because of their greater need for iron. For this reason, the enrichment of staple foods, especially those consumed in poor countries, is one of the top priorities in international agricultural and nutrition research. In rice-eating populations, iron deficiency anaemia is caused by insufficient dietary iron, absorption inhibitors such as phytate, and lack of enhancing factors for iron absorption such as ascorbic acid.

Although much is known about the uptake of iron and zinc in roots and transport of minerals to and from vegetative parts of the plant, some plants accumulate very little trace minerals in the grain (Holm *et al.* 2002). For example, in wheat only 20 per cent, and in rice just 5 per cent of the iron in leaves is transported to the grain. In cereals, much is stored in the husk and subsequently lost during milling and polishing. Thus, strategies to increase the iron content of cereals and grains face the challenge of targeting iron storage in a form and location in the plant where it will be bioavailable when consumed.

A significant breakthrough in improving the iron content of cereals was achieved by Ingo Potrykus and colleagues. One of the genes transferred to Golden Rice came from the common bean, *Phaseolus vulgaris*. This gene encoded for the iron storage protein, ferritin, and when expressed in the transgenic rice increased the iron content twofold. The bioavailability of iron in transgenic rice varieties containing ferritin was shown to be as good as ferrous sulphate, commonly used in iron supplements, as reflected in biochemical indices of iron status in iron-deficient laboratory rats (Murray-Kolb *et al.* 2002).

A different source of ferritin genes, soybean, was used in the transformation of rice to increase iron content (Goto *et al.* 1999). Researchers at the Central Research Institute Electric Power Industry, Japan, transferred the gene for ferritin from soybean into rice and confirmed the stable incorporation of the ferritin subunit in the rice seed. Iron content in the transgenic rice seeds was up to threefold greater than in non-transgenic control plants.

Others have used recombinant soybean ferritin under a different promotor to increase the iron content in wheat and rice. In this case, iron was significantly increased in vegetative tissues but not in seeds (Drakakaki *et al.* 2000). Thus, the experimental conditions, type of promotor used, mineral transport and storage in the plant, and other conditions have substantial effects on the outcome of genetic engineering experiments to increase mineral content.

Iron transport and uptake in plants is carefully regulated. This is because iron has low solubility and is toxic in excess. Recent studies have examined the function of iron transporter proteins in transgenic plants. These proteins have been shown to increase iron uptake into roots when iron is deficient (Eide *et al.* 1996). The iron transporter protein, IRT1, first isolated from *Arabidopsis thaliana,* also transports other metals such as zinc, manganese, lead, and cadmium; the latter two can be toxic. Researchers in the laboratory of Dr. Mary Lou Guerinot at Dartmouth College, Hanover, NH, have shown that slight changes in the amino acid composition of the transporter protein affects the selectivity of metals transported into the plant (Rogers *et al.* 2000). This finding introduces the possibility of engineering plants with the ability to take up desirable minerals while excluding toxic and undesirable ones.

A second iron transporter protein, IRT2, has also been identified in the roots of *Arabidopsis*. When the gene for IRT2 was incorporated into iron- and zinc-deficient plants, iron uptake was increased (Vert *et al.* 2001). Unlike the IRT1 transporter, however, IRT2 did not transport manganese and cadmium when it was expressed in yeast. This observation suggests ways in which selective genetic transformations might be used to enhance the uptake of some minerals while excluding others.

It should be noted that iron and zinc levels tend to be present together in many plants, although the average content of each differs. For example, in screening over 1,000 varieties of common beans, a nutritional staple in many countries, scientists at the International Institute of Tropical Agriculture, Nigeria, found that iron content averaged about 55 μg/g iron, but some varieties from Peru averaged more than 100 μg/g iron (Gregorio 2002).

Zinc content, averaged 35 μg/g. When varieties were selected for their iron content, higher zinc levels were obtained as well. These observations suggest that genetic modification of selected varieties to further increase iron content might boost zinc levels too.

As in beans, iron and zinc concentrations differ across varieties of rice. Aromatic rice tends to have the highest iron levels and several varieties have been successfully crossed with elite rice lines having excellent agronomic characteristics and grain qualities. These micronutrient traits were shown to be stable across different growing environments and could be crossed with high yielding varieties to improve the nutrient density. High iron rice developed from traditional breeding is currently being tested for iron bioavailability and effects on iron nutrition status in young women in the Philippines (World Bank 2000).

Results are not Yet Available

Another approach to improving the availability of iron for infants was reported by Suzuki and colleagues (2003) at the University of California, Davis. These investigators developed transgenic rice in cell culture that expressed the gene for human lactoferrin, a milk protein that binds iron. When they compared the recombinant lactoferrin with native human lactoferrin both proteins retained functional activity after mild heat treatment, high acidity, and *in vitro* digestion. Their findings suggest that recombinant lactoferrin grown in plant culture may be a functional alternative to human lactoferrin in infant formula and provide another way to improve iron availability during infancy.

Zinc

Zinc is one of several trace minerals that can be deficient in human diets, especially where meat is not consumed. Zinc deficiency is associated with impaired growth and reproduction, anorexia, immune disorders, and a variety of other symptoms. Zinc is also an important constituent of more than 100 enzymes. Absorption of zinc from cereals and grains can be impaired or blocked by the presence of some substances such as phytate.

Increasing the zinc content of cereals and grains, especially where soils are low in zinc, may be an effective way to improve human nutrition and at the same time increase plant yields.

Ramesh and colleagues (2004) in Australia, studied the effect on seed zinc content in barley (*Hordeum vulgare* cv Golden Promise) in plants transformed to increase the expression of zinc transporter enzymes. Multiple transgenic lines exhibited higher zinc and iron contents in their seeds compared with control plants (Ramesh *et al.* 2004). When grown under zinc deficient conditions, zinc uptake in the transgenic lines was higher in the short term compared with control plants. When zinc content was restored, uptake of zinc decreased in both transgenic and control plants, suggesting that the transporter proteins may be degraded when zinc is adequate. This study suggests that increasing the production zinc transporter proteins may be one approach to increasing the zinc content of cereals.

Selenium

Selenium is an essential trace mineral incorporated into plants from soil. Consumption of selenium has been linked to reduced risk of all cancers, but particularly those of the lung, colo-rectum and prostate (El-Bayoumy and Sinha, 2004, Combs 2004). Selenium is also important for specific enzymes and proteins in the brain and is necessary for proper immune function. However, selenium is toxic at levels only a little greater than those required in a healthy diet, so caution is warranted with supplementation and increased intakes. Areas where soils are deficient in selenium are well known and low to deficient intakes have been observed among human and animal populations in these regions.

Genetic engineering technology offers considerable potential for increasing the uptake of selenium from soils and incorporating the mineral into non-toxic compounds in the edible parts of plants. Plants genetically modified to absorb above average quantities of minerals could be used to improve human or animal nutrition. For example, a study from the University of California, Berkeley reported that genetically engineered *Arabidopsis thaliana* and Indian mustard (*Brassica juncea*) were able to incorporate more selenium from soil and convert it to non-toxic methylselenocysteine than wild type plants (LeDuc *et al.* 2004). Researchers at Purdue University, West Lafayette, In, also showed that plants not normally accumulating selenium, such as *Arabidopsis*, can be transformed to do so (Ellis *et al.* 2004). These studies demonstrate the feasibility of developing crop plants with improved ability to take up selenium from the soil and store it in a non-toxic form. Thus, selenium at appropriate concentrations would be safe for the plant and for human consumption.

Phytonutrients and Novel Substances

Intense research activity is being devoted to the identification and study of phytonutrients in plants with an eye to their ability to protect health, improve immune function, and reduce the risk of chronic diseases ranging from heart disease and cancer to age-related macular degeneration. Examples of phytonutrients include: phytoestrogens, polyphenols, and isothyocyanates. In spite of their cumbersome technical names, these various categories of substances appear to hold significant potential health benefits when consumed in modest amounts. Because they are widely distributed in fruits and vegetables, diets rich in these foods are likely to furnish generous amounts of many of these phytonutrients.

Unfortunately, there is insufficient scientific data from carefully controlled studies that adequately demonstrate safety and efficacy of substances with potential promise. For many substances—lycopene, isoflavones, resveratrol—to name a few, data appear promising, but are not consistent or conclusive. Extensive media and manufacturer publicity about many of these compounds generated expectations exceeding scientific justification. For these reasons, the enhancement of foods with particular phytonutrients usually lacks sufficient scientific grounding to justify the development of foods with enhanced levels. However, a few examples can be cited.

Isoflavones: Isoflavones are a type of phytoestrogen, so named because they bind to estrogen receptors and mimic some of the effects of the hormone estrogen. However, the biological effects of isoflavones differ markedly from estrogen and many are non-hormonal.

Soybeans are the richest food source of isoflavones, but isoflavones occur in other legumes such as broadbeans, and in many vegetarian ("meatless") foods made from soy products (USDA 2002). They have been linked with easing menopausal symptoms, improving bone health, reducing cardiovascular risk, and possibly reducing the risk of prostate cancer. The main soy isoflavones are genistein, diadzein, and glycitein. Their concentration in legumes is greatly affected by growing conditions and climate, and these variables could potentially override genetic modifications.

Currently, isoflavones are abundant and readily available, so genetic modifications to increase isoflavone content might not be expected to be a high priority. However, Yu and colleagues (2003) at the DuPont Company, Wilmington, DE, reported the application of genetic engineering techniques to increase the isoflavone content of soybeans. By activating genes in the phenylpropanoid pathway, diadzein levels increased and genistein levels fell.

By blocking the anthocyanins branch of this synthetic pathway, the investigators obtained higher concentrations of isoflavones. Thus, it is possible to increase the level of isoflavones in soybeans.

Phytosterols and Phytostanols: Phytosterols and their saturated derivatives, phytostanols, are plant sterols found in small quantities in vegetable oils. When consumed in sufficient amount, they are effective in reducing blood cholesterol levels. Sufficient data of their efficacy and safety exist that the FDA has permitting food manufacturers to claim a role for plant sterols or stanols or their esters in reducing the risk of coronary heart disease.

In 2003, researchers at Unilever Research, Netherlands, reported the generation of transgenic tobacco seeds with enhanced total seed sterol level (Harker *et al.* 2003).

Following the insertion of the gene for a key enzyme in sterol biosynthesis, total seed sterol levels increased by

2.4-fold. The additional sterol was present as fatty acid esters and several intermediate sterols accumulated. This group also developed transgenic tobacco carrying two genes involved in regulating carbon flux through the sterol biosynthesis pathway (Holmberg *et al.* 2003). The two transgenes increased total seed sterol content more than with either gene expressed singly (2.5-fold vs. 1.6-fold).

Researchers at Monsanto reported the application of genetic engineering to modify the ratio of phytosterols to phytostanols in rapeseed (*Brassic napus*) and soybean (*Glycine max*). Plants were transformed with a gene from yeast for the enzyme 3-hydroxysteroid oxidase (Venkatramesh *et al.* 2003). Seeds from both types of plants exhibited conversion of the major phytosterols to phytostanols and no other functionalities were affected. Several novel phytostanols were obtained as well. Because these substances are hydrogenated they would be expected to be more stable during food processing, yet still confer cholesterol-lowering benefits.

Most recently, Enfissi *et al.* (2005) reported the development of transgenic tomatoes with a 2.4-fold increase in phytosterols. Increases were greatest for phytoene and beta-carotene.

Such an alteration in a widely consumed food would potentially increase the consumption of these substances in a large share of the population.

Probiotics: The term "probiotics" refers to live microorganisms that have a health benefit when consumed in adequate amounts. They are usually bacteria selected from species found in the intestinal tract. Probiotic microorganisms may be concentrated and added directly to a food or added to a milk product in small amounts and allowed to grow. The most common foods having probiotic organisms are fermented dairy products such as yogurt containing *Lactobacillus acidophilus*.

Many health benefits have been attributed to probiotics, including resistance to infectious diseases, prevention of vaginitis, production of anti-microbial metabolites and nutraceuticals, immunomodulation, and others (Ahmed 2003). Foods containing probiotic bacteria are abundant in Japan and parts of Europe, but are less developed in the US.

Probiotics have been used as dietary supplements and oral agents for intestinal disorders.

A genetically engineered strain of *L. lactis* subsp. *diacetyllactis* is used as a buttermilk starter culture in the US (Renault 2002).

Probiotics have been used to treat inflammatory bowel disease by creating more hostfriendly gut flora. Selective use of probiotic bacteria can create an environment where stimulation of the immune system is restrained and intestinal inflammation reduced (Guarner *et al.* 2002). Some strains of *Lactobacilli* have prevented the development of colitis in genetically susceptible mice. Other genetically engineered bacteria have been used to secrete the anti-inflammatory cytokine IL-10 (Guarner *et al.* 2002). Considerable research is being devoted to the identification and effects of probiotic bacteria and their use as therapeutic agents, but few products have reached application in the US.

Genetically engineered probiotic bacteria have been used to overcome problems associated with more traditional technologies for developing such bacteria. In foods, genetically engineered bacteria have been used to improve the flavour and stability, or to block the formation of unwanted flavours. Genetic engineering should make it possible to strengthen the effects of existing bacterial strains and create new ones (Steidler 2003).

Anti-nutritional Substances

Many plants contain substances that inhibit nutrient uptake in various ways or are toxic themselves. They may interfere with intestinal cell function, reduce the ability to break down complex molecules such as proteins and starch, or may be toxic if consumed in sufficient quantity. Some of these substances are rendered harmless with cooking or processing, but others are resistant to digestion, heat treatment, or other forms of processing.

Although anti-nutrient substances reduce nutrient availability, many benefit human health and plants. For example, dietary fibre improves colon function and reduces plasma cholesterol levels; polyphenols and other substances have anti-carcinogenic properties.

Many of these substances are important in plant metabolism and provide resistance to environmental stress and pests. Further, some of the adverse effects on nutrient availability may diminish over time or with different levels of consumption, and this suggests that humans may be better able to adapt to these substances than was once thought (Welch 2002). The question of improving micronutrient availability by reducing anti-nutritional substances through plant breeding, biotechnology, and other means requires careful consideration of the complexities.

There are several ways of reducing anti-nutritional substances, including plant breeding, heat, processing, fermentation and drying, and genetic engineering. Examples of the various anti-nutritional substances that have been reduced by the application of biotechnology in different plants are described below.

Phytate: Phytic acid is a phosphorus storage compound found in the seeds of many edible crops, e.g., wheat, corn, barley, rice. Phytic acid forms salts (phytates) of potassium, magnesium, calcium, iron, zinc, and other minerals that cannot be absorbed. Phytic acid-containing foods bind minerals in the intestinal tract rendering them unavailable. When these minerals are limiting in the diet, the presence of phytic acid can contribute to mineral deficiencies, particularly in the case of iron and zinc. This is a particularly important consideration in the diets of women and children where legumes and cereals are staple foods. In animal nutrition, especially for poultry and swine, phosphorus may have to be supplied in a more available form to overcome the loss due to binding with phytic acid. An additional consequence is the production of high phosphorus animal waste with adverse environmental effects.

Lines of corn, barley, rice, and soybean with slightly different phytic acid characteristics have been used to develop varieties with reduced seed phytic acid (Raboy 2002). Reduction in phytate in the range of 50 per cent to 66 per cent has been achieved with these mutant lines. In soybeans and corn, 80 per cent reduction has been achieved. However, several hybrids developed with the mutant strains exhibited lower yields. It has now been shown that low phytate mutant corn is linked to the reduced expression of the enzyme myoinositol phosphate synthase, the first enzyme in the synthesis pathway for phytic acid (Shukla *et al.* 2004).

This finding was confirmed recently by Italian researchers who developed a mutant corn with 90 per cent less phytic acid and a 10-fold increase in seed-free phosphate (Pilu *et al.* 2003).

Proof that zinc absorption from low phytate corn compared with wild-type varieties was significantly greater was reported by Hambridge *et al.* (2004) who fed corn tortillas to six healthy adults. Zinc absorption from the 80 per cent phytate-reduced corn was three times greater than from the wild-type corn, 4.9 mg/day compared with 1.5 mg/day.

Although genetically engineered low phytate crops have not been commercialised, biotechnology has been used to express the enzyme phytase in plants (Chier *et al.* 2004). This enzyme allows animals to metabolise phytic acid and eliminates the need to supplement feed with phosphorous. Consequently, using animal feed engineered to produce phytase addresses many of the problems associated with high phytate levels in animal feed. Phytase has been successfully incorporated into soybean and wheat and is biologically active when the plants are used as animal feed (Brinch-Pedersen *et al.* 2000).

In a study of broiler chickens, consumption of transgenic soybeans containing phytase led to a 50 per cent reduction in phosphorus excretion compared with a diet supplemented with an intermediate level of non-phytate phosphorus (Denbow *et al.* 1998). Feeding the transgenic soybeans resulted in an 11 per cent greater reduction in phosphorus excretion than feeding with conventional soybeans to which the enzyme is added. Similarly, low-phytate corn and barley fed to broiler chicks resulted in 33 per cent lower phosphorus excretion compared with wild-type grain diets and reduced the need for supplemental phosphorus (Jang *et al.* 2003).

Transgenic phytase-containing wheat was developed by Holm and colleagues at the Danish Institute of Agricultural Sciences, Denmark (Brinch-Pedersen 2000, Holm *et al.* 2002).

Transgenic plants exhibited up to 4-fold higher phytase activity compared with wild-type seeds. However, a drawback of such a transformation for human consumption is that the enzyme is inactivated above 60°C and would be destroyed by cooking (Holm *et al.* 2002). It is possible that more heat-stable forms of the enzyme could be used, but when more heatstable enzymes were incorporated into rice, which was then cooked, only 8 per cent of the activity remained (Holm *et al.* 2002).

Transgenic rice expressing phytase derived from modified yeast genes has also been reported (Hamada *et al.* 2005). By selectively modifying the genes, the investigators were able to increase the enzyme activity above that of the original yeast gene.

Reducing the phytate content of plants, particularly soybean, has direct implications for human nutrition. For example, soy protein used in infant nutrition may limit mineral absorption because of its phytate content. To investigate this question, Davidsson and colleagues at the Swiss Federal Institute of Technology compared regular and dephytinised soy formula in nine infants 69 to 191 days old. Regular and dephytinised formula contained 300 and 6 mg phytic acid/kg liquid, respectively. The investigators reported that zinc absorption was significantly greater from dephytinised formula compared with regular formula, 22.6 per cent compared with 16.7 per cent absorption (Davidsson *et al.* 2004). Absorption of iron, manganese, copper and calcium did not differ between the two formulas. These findings suggest that use of dephytinised soy protein improves zinc absorption and can be recommended for infant foods.

Another approach to solving the phosphorous uptake problem in animal production was undertaken by researchers at the University of Guelph, Canada. In this case, swine were engineered to produce the enzyme phytase (Golovan *et al.* 2001). These pigs expressed the enzyme phytase in their saliva and exhibited complete phytate digestion, required no dietary inorganic phosphorus and excreted 75 per cent less phosphorus.

Gossypol: Cottonseed contains the polyphenolic compound gossypol, long known to be toxic to humans and animals. Interestingly, gossypol also appears to have anti-cancer properties towards several human prostate and breast cancer cell lines (Liu *et al.* 2002, Jiang *et al.* 2004). Martin and colleagues at Texas A&M University, College Station, TX, created a transgenic cotton (*Gossypium vitifolium*) using the antisense gene technology for a key enzyme in the synthesis of gossypol. Transformed cotton plants had up to 70 per cent less gossypol in their seeds compared with non-transformed plants (Martin *et al.* 2003). These findings suggest that biotechnology can be used to reduce the gossypol levels to render the seed oil more suitable for feed and food.

Cyanogens: Cassava (*Manihot esculenta* Crantz) produces various *cyanogenic glycosides* such as, linamarin, lotaustralin, and acetone cyanohydrin in its roots and leaves. These potentially toxic substances are only present in small amounts in "sweet" varieties of cassava, but are sufficiently abundant in "bitter" varieties to require removal (Padmaja 1996). Boiling and drying reduce these substances to safe levels in low cyanogen varieties, but those with higher levels require, soaking, grating or maceration, fermenting, and sundrying to reduce cyanogens adequately. The toxicity of these cyanogens is exacerbated by low protein intakes, a characteristic of countries where cassava is a staple.

In 2003, Drs. Siritunga and Sayre at the Ohio State

University, Columbus, OH, reported the development of transgenic cassava with a 99 per cent reduction in linamarin in the roots and between 60 per cent and 94 per cent reduction in leaves (Siritunga and Sayre 2003). These plants were transformed by inhibiting the expression of two genes that catalyse the first step in the synthesis of linamarin. The following year, this group reported that transgenic cassava roots expressing a different gene contained significantly less acetone cyanohydrin levels compared with wild-type plants (Siritunga *et al.* 2004). These accomplishments open the door to the development of cassava truly safe for human and animal consumption.

Steroidal Glycoalkaloids: Potatoes (*Solanum tuberosum* spp) contain potentially toxic steroidal glycoalkaloids, the best known of which is solanine. This substance is found in the green tissue of potato just under the skin. While these compounds protect the plant from pests, they reduce food quality and safety. Glycoalkaloids are synthesised from cholesterol.

Recently, Arnqvist *et al.* (2003) showed that the inhibition of plant sterol synthesis in transgenic potatoes reduced the synthesis of glycoalkaloids by 41 per cent in leaves and 63 per cent in tubers. Other investigators used antisense technology to create transgenic potatoes with up to 40 per cent less steroidal glycoalkaloids in the tubers (McCue *et al.* 2003). These studies indicate that substantial improvements in the reduction of steroidal glycoalkaloids in potatoes may be close at hand. It remains to be seen whether pest resistance in these transgenic potatoes is affected by the reduction in glycoalkaloids.

Mycotoxins in Corn: An important risk to the safety of grains (corn, wheat, barley), groundnuts (peanuts), tree nuts (almonds, walnuts, pistachios) and cottonseed is the production of toxins by fungi. Certain types of fungi—*Aspergillus flavus* and *Fusarium*—are notorious for their deadly products. The substances produced by these organisms cause disease in plants and potentially serious illness in people and animals consuming the infected crops. Aflatoxin, a particularly dangerous cancer-causing mycotoxin produced by the fungus *Aspergillus flavus*, is sometimes found in peanuts and corn. Fumonosins produced by Fusarium fungi are thought to be carcinogenic and harmful to the immune system. Agricultural practices and grain storage conditions help to minimise growth of these fungi, but weather also contributes to their development. Breeding crop varieties with increased resistance to fungi may reduce the production fungal toxins. Researchers at the US Department of Agriculture developed transgenic walnut trees that displayed increased resistance to aflatoxin synthesis (USDA 2004). Partial resistance to *Fusarium* disease was reported in transgenic wheat (Okubara *et al.* 2002) and in transgenic bacillus thuringiensis (Bt) corn (Bakan *et al.* 2002). The potential for genetic engineering strategies to reduce the production of fungal toxins in food crops has been discussed by Duvic (2001) and Munkvold (2003).

Oxalates: Oxalic acid is present in spinach, tomato, groundnut, soybean, and chick pea.

It binds several minerals including calcium and prevents their absorption. Scientists at the National Centre for Plant Genome Research (NCPGR), India, seek to reduce the oxalic acid content in these foods through the transfer of the gene for oxalate decarboxylase (OXDC), the enzyme that degrades oxalic acid. Genetically engineered tomatoes bearing the OXDC gene have been successfully grown and are currently undergoing field and biosafety tests (ISAAA, 2004).

Allergens

Food allergy, although relatively rare, can provoke severe, sometimes fatal, responses in susceptible people. Specific food proteins can trigger the immune system and provoke an allergic response. Risk of such reaction is greatest in the first two years of life, and by age five about 80 per cent of allergic infants will lose their food allergies.

Eight types of foods account for nearly 90 per cent of all

food allergies. These are: milk, eggs, fish, crustacea, wheat, peanuts, tree nuts, and soy. People allergic to one type of food are frequently, but not always, allergic to others. Many proteins associated with food allergies in these foods have been identified, but many remain unknown.

Soybean: In people with soy allergy, as many as 28 proteins may bind with IgE, a type of antibody involved in allergic responses, suggesting that many soy proteins have allergenic potential. More than half of soybean allergic reactions are attributable to a single protein, known as P34 (Cordle 2004, Wilson *et al.* 2005).

Scientists at USDA and the Donald Danforth Plant Science Centre have succeeded in silencing the gene for P34 and created soybeans without this protein (Herman *et al.* 2003). However, two other proteins that trigger allergic reactions may have to be removed before soybeans could be sold as hypoallergenic. Dr. Anthony Kinney, a researcher involved in the project, commented that removing the other proteins should not be difficult because wild species lacking the genes for the other proteins are already known. Careful plant breeding with the genetically modified soybeans may be able to produce the hypoallergenic soybeans.

Other genetic engineering strategies that alter the composition of the allergenic proteins may render the proteins harmless to sensitive people. The ability to eliminate the most hazardous allergens in soy would have substantial benefit for infant formula feeding and for those who are allergic to soy.

Peanut: Allergy to peanut proteins can be fatal. The major allergens in peanut have been identified as Ar h1, Ar h2, and Ar h3, and their genes have been isolated. The protein Ar h2, a Kunitz trypsin inhibitor, is believed to be the most potent of these three (Koppelman *et al.* 2004). Recombinant versions of these allergens produced in bacteria were heat-killed and used in a vaccine (Li *et al.* 2003). The vaccine was given in three different doses to allergen-susceptible mice. Animals were challenged after 2, 6, and 10 weeks. Treated mice produced no anaphylactic or histamine response when challenged. Animals given the medium and high doses remained protected for 10 weeks. This particularly encouraging research suggests that protection against peanut allergy may be in the foreseeable future.

Rice: Rice is generally consider a hypoallergenic food, and for that reason is one of the solid foods first introduced to infants. Nonetheless, some people, particularly in Japan, are allergic to rice proteins. The major allergen(s) in rice have been identified and hypoallergenic varieties of rice developed using antisense genetic engineering techniques to suppress the synthesis of the predominant allergen (Nakamuro and Matsuda 1996).

Allergen content in the transgenic seeds was reduced from about 300 micrograms/seed to about 60 to 70 micrograms/seed. However, when the hypoallergenic rice was tested in sensitive patients, not all allergenic potential had been eliminated. Extremely sensitive patients were also sensitive to other proteins, so that a single genetic transformation was insufficient to overcome their allergic responses. Europa Bio (2002) reported that several laboratories in Japan are working to develop hypoallergenic rice, but efforts have provided only partial success, as it is not possible to eliminate all allergens.

Shrimp: Allergic reactions to shrimp and other crustacea (lobster, crayfish) are among the most common food allergies. People with hypersensitivity responses to crustacea may also be allergic to molluscs, and some arthropods such as house dust mites and cockroaches. To date there have been no ways to overcome these allergies. The major allergen in shrimp is the muscle protein tropomyosin, known as Pen a 1. Recent analysis of this allergen revealed five IgE-binding regions or epitopes (Ayuso et.al. 2002). These regions contained 15 to 38 amino acids from whose sequence the corresponding gene sequence can be determined. Dr. Samuel Lehrer of Tulane University, New Orleans, LA, has located the gene sequence that encodes these epitopes and has suggested that by altering these epitopes by a single amino acid, IgE binding could be halted. This work suggests that safe recombinant allergens could be synthesised for immunotherapy of those who are allergic to shrimp and related substances.

Miscellany

Beer and Wine: Applications of biotechnology are finding their way into brewing and wine-making. One application in grapevines (*Vitis vinifera* L.) has been increased content of the anti-oxidant resveratrol. This substance is of special interest in human health for the reduction of oxidised lipids. It is also thought to improve plant resistance to fungal disease. Gonzalez-Candelas *et al.* (2000) also reported increased resveratrol in wine through the use of transgenic yeast. In a different approach, Giorcelli *et al.* (2005) developed transgenic grapevines using the gene for stilbene synthase, an enzyme that leads to the production of resveratrol. The plants increased their production of resveratrol, but showed no improved resistance to a leading fungal pathogen.

Transgenic yeast also has potential for modifying the flavour of beer (Vanderhaegen *et al.* 2003) and sake (Aritomi *et al.* 2004). Genetic engineering also holds considerable potential for the production and regulation of diverse flavours and aromatic substances, as recently described (Dudareva and Negre 2005).

Decaffeinated Coffee

Japanese scientists have succeeded in silencing the gene responsible for caffeine production in coffee plants (Jameel 2003, Ogita *et al.* 2003). Caffeine content in the transgenic plants was reduced by 70 per cent. The researchers aim to modify Arabica coffee plants, the most popular coffee grown. Towards this end, Ogita *et al.* (2004) reported additional progress on modifying the pathways in Arabica and canephora coffee varieties which resulted in caffeine reductions ranging from 30 per cent to 50 per cent. This genetic modification would eliminate the need for decaffeination processing.

3

Biotechnology and Bioincubators

Biotechnology employs biological systems of microbial, plant, or animal origin, or specific compounds derived from these systems for the production of goods, including foods. Biotechnology also uses living organisms at the cellular or molecular level for agricultural, medical, or industrial purposes. The publication of the human genome sequence in 2001 brought biotechnology dramatically before the public as a leading-edge scientific endeavour. Basic biotechnology is however thousands of years old. The basics of biotechnology were known to Indians and others on the preparation of curd, "Soma Rasa" (Alcohol) including fermentation since 3000-2500 BC. Baking and brewing also involved use of biotechnology. Later discovery of antibiotics opened a new era in biotechnology. The Gene technology and biotechnology is now considered as the application of scientific and engineering principles to the processing of materials using biological organisms and/or processes, etc. to produce goods and services for mankind. (Abelson, 1984, Marx, J.L. (ed), 1989, Eisenberg, 1992).

Figure- DEVELOPMENT OF MODERN AND NEW BIOLOGICAL SCIENCES

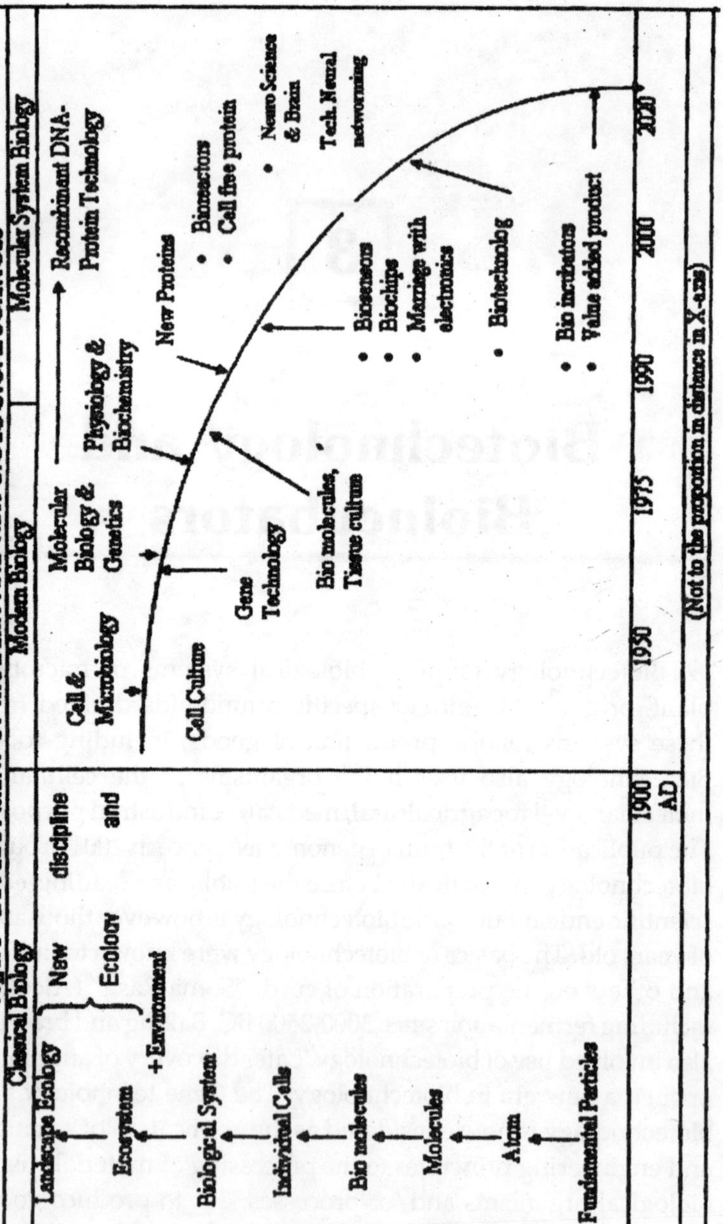

(Not to the proportion in distance in X-axis)

Modifying plant and animal genes are not new. Typically this was done through hybridisation in which two related plants/animals were cross-fertilized and the resulting offspring has characteristics of both parents. Breeders then selected and reproduced the offspring that had the human-desired traits. Hybrid maize was an early success of applying genetic theory to food production. Hybrid farm animals were produced in this way. During the 1930s, hybrids replaced open pollinated varieties of maize in USA and by the 1960s virtually all maize crops planted in US were hybrids. In 1997, the US maize yields average of increased to an eight tonnes/hectare, compared with one tonne/hectare in 1930.

Yeast has been a component of baking and fermenting throughout recorded history. Vaccination against the smallpox virus was introduced in the 18th century, long before the details of cell structure and action were elucidated. But it was James Watson and Francis Crick's discovery in 1953 of the structure of DNA, the molecule that carries genetic information, gave firm footing to Molecular genetics and biotechnology.

In 1953 this development ushered in the modern era of biotechnology. Since then, the science of genetics and its technological applications have advanced rapidly. In 1973, the first alteration of a DNA molecule could be done and the process is now referred to as recombinant DNA technology. In 1982, the US Food and Drug Administration approved the first drug, human insulin, produced in genetically modified bacteria. In 1989, and 1990 cotton and corn were respectively genetically modified to protect them against insect pests.

The first animal cloned from an adult cell of Dolly, the sheep, in 1997. Number of biotechnology drugs and vaccines are used today in the industrially developed countries, and agricultural applications of biotechnology are phenomenal. Industrial uses of biotechnology are fast growing. Biotechnology is used to address remediation of environmental pollution and remedial of environmental degradation and resource recovery.

Biotechnology Techniques

Understanding of the biological, chemical and biophysical aspects of a living cell (system) is the basic requirement for the development of biotechnology techniques:

(a) Cell culture technology is the growing of cells outside the living organism in which they develop naturally. Applications of biotechnology include growing cells on which to test new medicines, growing cells to replace dead or malfunctioning cells in human organs, and mass production of natural substances of medicinal value.

(b) Cloning is the reproduction of molecules, cells, plants or animals that are genetically identical to their source. Cloning gained popularity in 1997 when scientists cloned a sheep from an adult sheep cell. Before that, animal cloning had been done with embryo cells. Monoclonal antibody production and much of cell culture technology are based on cloning. Cloning is also used in livestock breeding, and modification of agriculturally important plants. (Willadsen, 1989, Zanwar, 1989, Pursel *et al.*,1989).

(c) Genetic engineering or recombinant DNA technology is the insertion of DNA from one organism into the genetic material of another organism. This technology is a more specific and direct approach to the same ends as selective breeding. (Hammer, *et al.*, 1985, Glover, 1985, Rigby, 1987).

Other uses of this technology include production of medicines and vaccines, treatment of genetic diseases, and nutritional enhancement of foods. Monoclonal antibody technology develops antibodies from cloned cells that can be used to identify and treat antigens that infect humans. Because antibodies are very specific in their action, monoclonal antibody technology encompasses an extensive field of research. One of the more important applications of this technology is cancer treatment and development of vaccines, such as the biotech vaccine against hepatitis B. Monoclonal antibody technology is also used to locate environmental pollutants and to detect harmful microorganism in food. (Golding, 1983).

(d) Proteins are the chemical substances through which much genetic and cellular activities occur. Research efforts are made to understand and manipulate proteins. Currently, there are several biotech drugs based on protein chemistry for treatment of diseases like anaemia, cystic fibrosis, haemophilia, leukaemia, and some cancers. Besides their functions in living organism, enzymes are also used as biocatalysts to improve the efficiency of production processes for chemicals, pharmaceuticals, pulp and paper, textiles, food, and animal feeds. (De Gardo *et al.* 1989, Hodgson, 1991).

(e) Modern techniques in plant biotechnology: Techniques of modern plant biotechnology involve directed insertion of DNA into plant cells via agrobacterium or accelerated particle-mediated transformation. Following cell transformation, tissue culture techniques facilitate the development of fertile plants containing unique traits that are the objective of the transformation process. One of the success in plant biotechnology is a number of plant varieties capable of synthesising an insecticide protein encoded by gene isolated from bacterium, *Bacillus thurningiensis* (Bt). After ingestion by the susceptible insect, the protein is proteolytically metabolised to an active form that disrupts ion transport channels of cell membrances in the insect intestine. This leads to the death of the insects that destroy crop plants. (Goodman, et.al. 1987, Grierson, 1991).

(f) Genes known to code proteins resistant to water stress and drought and other stress factors in some bacteria and plants, have been successfully transferred to some crops, cash crops and other commercially important plants. For example transgenic tobacco and rice plants have been produced by transferring the gene P5CS from moth bean. Transgenic soybean tolerant to drought and water stress has also been produced. The gene, SacB found in the bacterium, *Bacillus subtilis* encodes for the enzyme Levan Sucrase, which mediates in the fructan (Polyfructose molecules) synthesis. This gene was transferred to tobacco to make it more resistant to drought. Transgenic plants have been genetically engineered by transferring genes from bacteria *(Serratia marcescens)* fungi *(Tricoderma harziannum)* and some plants to with stand fungal and bacterial pathogens. There are many such success stories of producing transgenic plants. Genes encoding anti-microbial proteins have been transferred to potato and many food plants. (á-Thionin gene, alf AFP gene, etc.). Another example is Biofertilizers which belong to two categories of microorganisms, (i) nitrogen fixing and (ii) phosphate solubilising.

Genetic engineering of crop plants by transferring nif genes from nitrogen fixing microorganisms have been done. Nif genes encode for nitrogenase enzyme and some protein factors, which catalyse the formation of ammonia from atmospheric nitrogen. Considerable researches are being carried out in this line. Another line of research in Environmental biotechnology is production of Biopesticides. *Bacillus thuringiensis* have a number of different strains like B. thuringiensis kurstaki, B. thuringiensis israelensis, B. thuringiensis tenebrionis, etc. and these strains are toxic to different groups/species of insects.

Using genetic engineering techniques, specific genes in these strains of bacteria are transferred to other field suitable bacteria like *Escherichia coli* or *Psendomonas fluorescens* and then field applications of these bacteria are done to protect crops from insect pests.

Agriculture Productivity

Great success has been achieved in increasing agriculture productivity to fulfil human needs of food and fibre during the second half of the 20th century. On the basis of the current population growth rate of 1.4 per cent per year, the world population will increase from the current level of six billion to approximately nine to twelve billion in 50 years. Providing food to this population will require enormous increase in agricultural production. Based on the minimum protein requirement for humans (0.8 g/kg body weight/day), it is estimated that for each additional billion people added to the planet, there will be an additional 14.5 million tonnes of quality protein requirement per year. In view of this:

i. Increasing the nutritional composition of plant foods should be a priority. The World Health Organisation recently reported that more than three billion people worldwide are malnourished. In addition, macronutrient and micronutrient deficiencies are widespread. Approximately 250 million children are at risk of vitamin A deficiency, which may lead to eye damage or blindness. This is a problem in India. Two billion people worldwide are at risk of iron deficiency, which is the most common nutritional disorder. It impairs immunity and adversely affects the physical and mental capacities of people of all ages. (Swaminathan, (ed), 1991, Swaminathan and Jana (eds), 1992).

Genetically Engineered Foods

One of the most exciting development in improving nutritional values of agricultural products using recombinant DNA techniques is the introduction of genes into rice that results in the production of β-carotene in the rice grain, β-carotene is a vitamin A precursor and a pigment required for photosynthesis. It is synthesised in the green tissues of all plants, including rice; but it is usually not present in non-photosynthetic tissues.

Traditional plant breeding has given us some plants that produce β-carotene in non-photosynthetic tissues such as roots of carrots. To get cells of the grain to produce β-carotene, three genes were inserted into the rice genome. Two were from daffodil and one from bacterium, *Erwina uredovora.* They specifically code for key enzymes of β-carotene synthesis. The grain of the transgenic rice has a light golden-yellow colour because of its β-carotene content. This rice particularly offers another food product to improve nutrition for the billions of people in developing countries including India where vitamin A deficiency is prevalent and where rice is a major food. Tomato fruit and its products—juices, soups, and sauces are the major dietary sources of β-carotene and the principal sources of lycopene, an immediate precursors of β-carotene in tomato. Like β-carotene, lycopene has received attention for the association between tomato consumption and improvement of human health, for example, reducing the risk of certain cancers. Lycopene is converted from its precursors phytoene by phytoene desaturase enzyme. Introducing a bacterial carotenoid gene (crtl) to tomato plants results in the conversion of phytoene to lycopene. Expression of this gene in plants results in a threefold increase in β-carotene in tomato fruit. Tomato eating should be encouraged and its availability to common people should always be looked into and its price should be affordable.

Other Agricultural Uses

Biotechnology has become an important tool of plant agricultural produce.

i. The most important use of genetic modification in plant agriculture is herbicide tolerance. In this process, the genetic composition of plants is altered to make them resistant to damage from the chemical herbicides used to kill weeds in the agricultural fields. In this process, crop losses from herbicides are reduced and yields are increased. When plants are made resistant to more lethal herbicides, fewer applications of herbicide can also be used, reducing both farmer's production costs and environmental damage.

ii. The second major use of genetically modified crop is insect resistance. This process involves taking genetic material from naturally occurring organisms that are lethal to insects and inserting it into plants. When the genetic insecticide from a naturally occurring bacterium is inserted in the genetic make up of plants, the insects that feed on them are killed before they damage the plant. This reduces the requirement of chemical insecticides, thereby protecting crops more efficiently and reducing the threat of poisoning animals and humans.

iii. Another recent advance is the use of genomics-based approach to increase á-tocopherol content in oil-seeds. Vitamin E is an essential nutrient and is a lipid-soluble phenolic compound synthesised in higher plants and some photosynthetic bacteria. Vitamin E refers to a family of related compounds called tocopherols. There are four types of tocopherols synthesised in plants: a, β and ë, and ä tocopherols, á-tocopherol is found in relatively low levels in all major oilseed crops, whereas its immediate precursor, ë-tocopherol methyl transferase (ë-TMT), is limiting in the oilseeds. Over-expression of the ë-TMT gene in the seeds of Arabidopsis results in an 80-fold increase in á-tocopherol without altering the total tocopherol content; this suggests the increase is due to the conversion of the large pool of ë-tocopherol. *Arabidopsis* is a small flowering plant that is widely used as a model organism for basic research in genetics and molecular biology, and it is very similar to the major oilseeds in which TMT is limiting. Therefore, it appears possible to convert a large pool of ë-tocopherol in oilseeds, including soyabean, to á-tocopherol by over-expressing the ë-TMT gene.

The safety issues surrounding foods from genetically engineered crops are key to their acceptance. Consumers seek reassurance about the safety of the foods they eat in terms of their 'immediate and long-term' health effects.

There are government agencies with regulatory authorities on genetically engineered products in different countries. The ultimate goals of these regulatory agencies are to ensure the safety of any products that will be used by consumers and to monitor and protect the environment. India should develop a comprehensive plan for testing of genetically modified foods for safety as it is done in USA and EU. There is concern that genetically engineered foods may introduce allergens. It is different from conventionally produced foods. In fact, it is feasible to eliminate possible allergens from food sources by using modern biotechnology.

Biotech Drugs

Biotech drugs have been introduced at an increasing rate, especially since the mid-1990s. A number of biotech drug products and vaccines available in the market. Current biotech medicines include important treatments for anaemia, cystic fibrosis, growth deficiency, haemophilia, hepatitis, transplant rejection, and leukaemia and other cancers. Biotech products are also used for several diagnostic procedures.

Hundreds of biotech medicines are undergoing clinical trials. Most new biotech drugs currently being tested are intended for cancer treatment. Other therapeutic categories with a large number of biotech drugs in the clinical trial phase are for infectious diseases, neurological disorders, heart disease, and respiratory diseases.

Microorganisms/specific animal species or processes associated with them are used for waste land reclamation, organic waste management, pollution control, metal recovery from ores and tailings, etc. (Singh, ed. 1992, Dash, 1993 1998, 1999, Dash *et al.* 1986, Kale, 1998, CPCB, 1999, 2001). Besides, environmental biotechnology is used for generation of electricity, biogas, lipid biofuel, alcohol, hydrogen, etc. (SPAN, 2006).

Environmental Biotechnology can be combined with conventional metallurgy called Biohydrometallurgy or Microbial mining or Bioleaching. In this process metals can be recovered from sulphide iron ores and tailings using the bacteria, *Thiobacillus ferroxidans, Thiobacillus thioxidans, Thiobacillus acidophilus, Thiobacillus organoparus, Ferrobacillus sulphooxidans.*

The general reaction for indirect microbial mining is:

Bacterial

(i) Metal sulphide

$$(MS) + Fe_2 (SO_4)_3 \xrightarrow[\text{Intervention}]{} MSO_4 + 2\,FeSO_4 + S$$

(ii) $$4\,FeSO_4 + O_2 + 2H_2SO_4 \xrightarrow[\text{Thiobacillus}]{} 2Fe_2(SO_4)_3 + H_2O$$ ferrooxidans

Presence of the bacteria in the acidic medium (at pH 2.0-3.0), the oxidation of Fe^{2++} is about $10^5 + 010^6$ times faster than chemical oxidation process.

Sulphur generated in the process is oxidised to H_2SO_4 by the bacteria (T.ferrooxidans and T. thiooxidans). The acid makes the medium acidic for faster bioleaching.

Direct bacterial leaching involves direct constant contact between the sulphide containing mineral and the bacterial species. Bacterial enzymes are involved in the process and ultimately:

$$\text{Metal Sulphide (MS)} + 2O_2 \longrightarrow MSO_4$$

Thiobadllus sp. are tolerant to Copper, Aluminium and Zinc along with Iron ores metals are recovered by bacterial mining and the general flowchart.

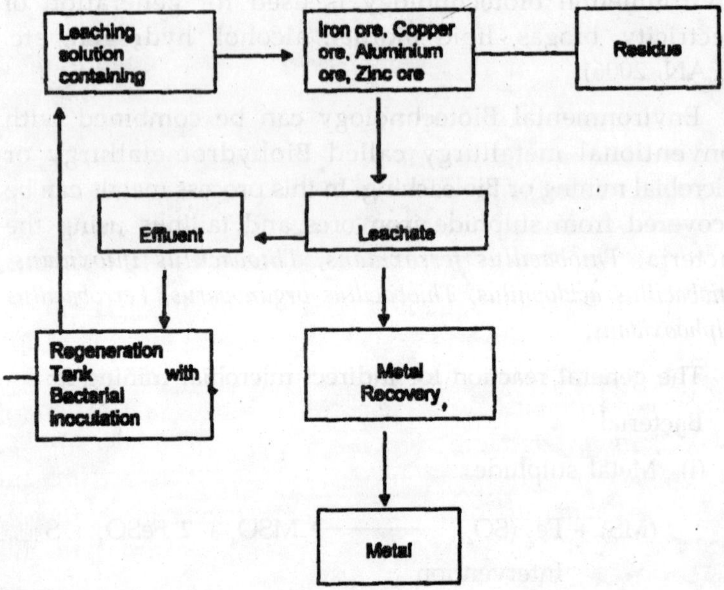

Bioincubators

Biotech firms traditionally have focused on Therapeutics like vaccine development, diagnostics like making kits for testing various diseases, and drug discovery. These days with the Human Genome being sequenced, 'Genomics' (The scientific study of genes and their role in an organism's structure, growth, health, disease (and/or resistance to disease, etc.), 'Functional genomics' (Study of, or discovery of, what traits/ functions are conferred to an organism by given (gene) sequences. Typically, functional genomic study follows after discovery of gene sequences found via structural genomics study) and Proteomics' (study and cataloguing of proteins, their structure and interactions and properties) have become the buzz words.

Stem cells and their use in therapeutics is big area here and so is a new technology (8-10 years old) called RNA Interference (using antisense RNA to block or knockdown specific gene expression). These are being routinely used to test for therapies against tumours and cancers.

Another area is 'chemical-genetics' where scientists use an array of organic small molecules to screen for biologically relevant function using a fast and high throughout breeding vertebrate like fish or frogs. For example various tadpoles are incubated in different chemicals and their subsequent phenotype is observed. Say on exposure to chemical X, tadpoles show inhibition of migration of pigment cells, or formation of another head. Then by looking at genes which cause similar phenotype, chemical X and its interaction with specific genes involved in head formation or pigment cell migration is further elucidated (Dash, 2005).

Like in Ecology, another very hot topic in molecular biology/biotech is 'System-Biology' at human body level where scientists are trying to identify not only individual genes but gene networks and every other component involved in the network(s), thus connecting RNA, DNA, genes, proteins and elucidating how they interact with each other to create breathing, blood pumping, disease fighting, food processing, problem solving human (Dash, 1999). Of course it is daunting since as you identify networks you start to look with a non-reductionist approach. Modelling gene networks and guessing certain outcomes at hypothetical physiological condition is also a big research thrust.

Bio-informatics and data mining is important (using computers and softwares to create information database that permit analysis of genomes, protein sequences, biomolecules and other organic matters). 'Microarray technology' (where one studies total gene expression patterns in response to certain drugs, certain diseases and so forth) is also hot since 10 years or so.

Bio-safety

A strong Bio-safety regulation is also required especially for storage and disposal of transgenic organism. Bio-safety regulation is very stringent in UK. Some Molecular Biologists routinely making transgenic frogs (Dash, 2006). And there is always a risk of these organism leaking into the wild and then contaminating native species with their transgene. They are used to make frogs which are transgenic for a protein called GFP (green fluorescent protein), so when you use this protein and if you see the frogs using UV then they glowed green. The scientists have to be very careful in handling and storage of these frogs and so were others who were making transgenic plants, transforming bacteria and viruses. Risk evaluation and control is a must for any Biotech park (Bioincubators).

Many Biotech parks have dedicated firms producing biologically active substances like enzymes, antibodies and other products which are routinely needed in molecular biology experiments.

Biofuels (Jatropha sp.) such as bioetahnol production and conversion of agri-waste as also quite popular and huge amount of money is invested on research.

All Biotech firms here have an in-house technology transfer personnel who can be patent lawyers or "idea sellers" some of them like to be called and they produce value added products.

There is new romance with biotech incubators. It would be great to see them pump out new companies such as Amgen, but almost by definition, most may fall and may lose tremendous amount of money in the process. And long before we carry around personal genomic data on credit cards, local governments are likely to face a revolt by taxpayers who believe they have been sold a false bill of goods. This threat is there and careful planning is required.

That's because biotech companies, particular those focused on drug discovery, are more delicate than orchids when it comes to the environment they need to survive. Building a successful biotech cluster takes smart ideas, smart people and smart money. And even when these things are in place, some plain good luck is needed.

Not all smarts can easily be found together. A 2002 study

by the brookings Institution {J. Cortright, H. Mayer, "Signs of Life: The Growth of Biotechnology Centres in the US" The Brookings Institution Centre on Urban and Metropolitan Policy (2002)} found that NIH grants in the life sciences have over the last several years been going to researches in increasingly diverse locations. Yet venture capital (VCs) funding in biotech, and the rare instances of commercial success, is actually becoming more concentrated. Just two areas, Boston and San Francisco, captured more than half the venture capital that went to biopharmaceuticals between 1995 and 2001. These examples are to be carefully analysed before taking policy decisions on establishing biotech firms.

Conclusion

Understanding of Molecular biology, especially Molecular genetics and molecular ecology is essential for the development of biotechnology and environmental biotechnology methods. Biotechnology, its principles and methods are the basics of biochemical engineering, sciences, which will be extensively used for human welfare.

Biotech incubators (Biotech Parks) should be built on a different model than other technology incubators. Given the higher barriers to entry, expectations need to be more modest. That could mean smaller parks and higher prerequisites for getting a space say, a certain level of government grant money or existing private investment.

That way, there will be less wasted on companies that ultimately can't pay the rent. The state biotech money could be spent improving existing infrastructure rather than trying to entice something that doesn't yet exist. For example, Minnesota in USA, eager for more biotech growth, is considering a $100 million investment in a genomics research facility at the Mayo Clinic.

The research that would come out of such an institution might be of great public benefit, and as a side effect may spur more private-sector entrepreneurship than a government programme can foster. This should be an eye-opener for developing countries. Investment in molecular biology, molecular genetics, and biotechnology research in Universities and Research institutes with time bound and goal oriented programme will be very much rewarding.

Since many developing countries including India are biodiversity-rich, development of Herbal gardens with the goal of producing value added products will be commercially profitable. This should be encouraged. Environmental biotechnology methods show promise for use as a tool to solve environmental problems in developing countries.

4

Bioremediation

Approaches towards Better Environment

Biotechnolgy is the application of biological organisms to technical, industrial and agricultural processes. This often involves the use of "novel" microbes, which have been altered or manipulated by scientists through techniques of genetic engineering, and the employment of these techniques in a wide range of economically important endeavours. Environmental biotechnology encompasses microbial intervention including bioaugmentation, biocontrol and bioremediation to have sustainability and quality in the output without use of antibiotics and chemicals.

Bioremediation means Biology "Remediate" To solve a problem, i.e. Bio-Remediate= to use biological organisms to solve an environmental problem. The term environmental biotechnology has a certain air of modernity when in fact it has a long history of use considering the underlying principles and not the appellation. However, as part of its complex meaning, there is a new definition and purpose in this discipline with regard to bioremediation.

Concept of Bioremediation

Bioremediation is the collective range of clean up methods by using natural microorganism (such as bacteria, plant, Fungi, etc.) to degrade hazardous organic contaminants or convert hazardous inorganic contaminants to environmentally less toxic or non-toxic compounds of safe levels in soils, subsurface materials, water, sludges, and residues. Bioremediation is a concept which includes all those processes and actions that take place in biotransformation of an environment, already altered by contaminants, to its original position (Thassitou & Arvanitoyannis, 2001).

Why Bioremediation?

Industrial revolution and anthropogenic activities have resulted in large quantities of chemicals being eventually released into the environment, either deliberately for agricultural and/or industrial purposes or accidentally by mishandling of chemicals, adversely affecting environmental qualities.Bioremediation have the potential to reverse this trend through environmentally compatible biological means.

The restricted or no use of antibiotics is encouraged because of the development of resistant bacteria and associated health cost. It is reported that billions of dollar is spent in US and Canada for tackling the resistant bacteria and associated higher antibiotic doses. There is a greater socio-economic justification like

1. In most cases a permanent solution to the contamination problem and it is the least expensive option.
2. Relatively non-intrusive.
3. Attractive alternative and/or supplement to more conventional clean-up technologies.
4. Safety for site workers as only minimum exposure to the contaminant and long-term protection of public health.
5. Environment-friendly "green technologies" and Effective.

6. Socio-economic development using natural resources to protect the environmental health.
7. In the era of judicial activism and environmental concern, bioremediation serves as the environmentally safe technology.

Principle of Bioremediation

Bioremediation is based on the principle that organisms are capable to take in things from the environment and use it to enhance their growth and metabolism. Bacteria, Protista, and fungi are well known for degrading complex molecules and transform the product into part of their metabolism.

The Process

Microbes releases enzyme to break down the contaminant into digestible pieces and the contaminant of organic substances is ingested and digested as food along with other energy source by the cell.

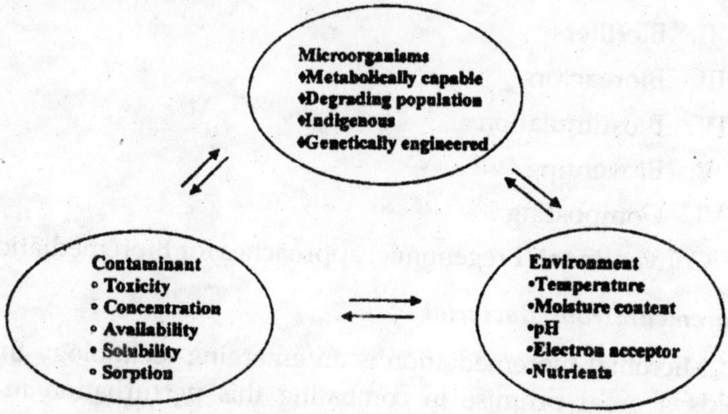

For our convenience we have classified Bioremediation into two major approaches: Conventional (Pre-genomic) and Genomic approaches. Now with lots of genomic information about potential microbe for remediation pouring hi through molecular approaches, it is possible to characterise and engineer novel organism to carry out bioremediation.

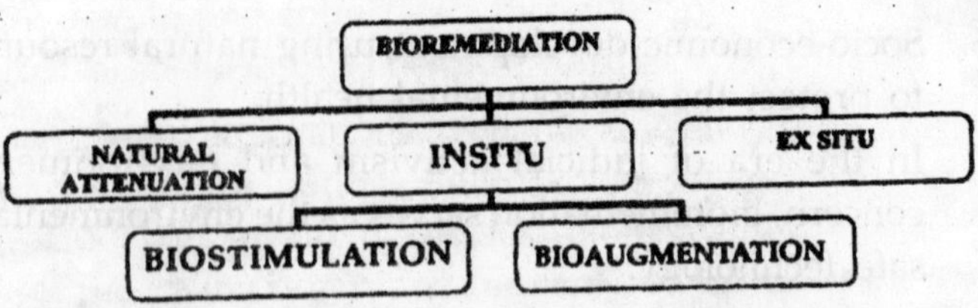

Bioremediation technologies are classified as either *in situ* or *ex situ* as shown in the above diagram. In situ technologies involve the use of organisms or enzymes to remove pollutants in the location that is polluted. *Ex situ* technology involves shifting the contaminated material to an area where it can be treated using bioremediation. Again, bioremediation could be natural when relying on naturally occurring microbial and plant processes or accelerated by addition of genetically modified organisms (GMOs).

The technologies which uses bioremediation principles includes:

I. Bioaugmentation

II. Biofilters

III. Bioreactors

IV. Biostimulation

V. Bioventing

VI. Composting

Conventional (Pregenomic) approaches for Bioremediation

Bioremediation: Bacterial

Microbial bioremediation is an emerging technology that holds a great promise in combating this perturbation in a environmentally compatible and sustainable way.

- The use of bacteria to degrade pesticides. Scientists have identified several bacteria that could be used to clean up pesticide residues in soil
- Bacteria used as detectors of pollution and monitors of toxic waste. These bacterial biosensors, such as modified

luminescent bacteria, can be used to measure the toxicity levels in samples of soils and water

- The possibility of using genetically modified (GM) plants and microorganisms as a means of remediating persistent pollutants, such as toxic explosive residues
- Huge numbers of bacteria exist naturally in contaminated environments. Some of those bacteria slowly break down different types of waste while some use these wastes as source of nutrients. These bacteria can be used to remediate these xenobiotic compounds.

In many cases, contaminants such as petroleum hydrocarbons serve as sources of organic carbon and electron donors (assimilation). In other metabolic processes, contaminants such as trichloroethylene may serve as electron acceptors (reductive dehalogenation), or may be transformed by processes that offer no added benefit to microbes from the reactions (co-oxidation). These processes occur within a wide range of hydrogeologic settings, biogeochemical interactions, microorganisms, and contaminants.

Microbes in Bioremediation

In case of the organic waste microbes like *Bacillus, Pseudomonas, Acinetobacter, Cellulomonas, Rhodoseudomonas, Nitrosomonas* and *Nitrobacter* are known to help in mineralisation of the organic wastes (Thomas *et al.*, 1992). Environment influence the bacterial activities-psychrotolerant bacterium that can be active at low temperatures may be a possible way to enhance heterotrophic activities during the winter season (Fukami *et al.*, 1999). Again the cellular and molecular immunomodulation in aquatic animal by the probiotic bacteria is also influenced by the environmental factors like temperature (Panigrahi, *et al.*, 2005,06)

The nature of Pollutants also influenes to the extent the microbial bioremediation can be effective The nature of pollutant can be of the following types Organic Pollutants: Examples, petroleum products and by-products, chlorinated organics (PCBs, etc.), nitro-aromatic compounds (TNT, etc.), pesticide residues, etc. Inorganic Pollutants: Heavy metals like Cd, Cu, Co, Pb, etc., pollutants like fly ash Soil Pollutants

Organic Microbial Intervention

Microbial processes employed in remediation strategies aim at the removal of excessive organic input through the mobilisation of elements (e.g., C, N, P) in the accumulation areas (Fabiano *et al.*, 2003). A conceptual model for application of bioremediation to face the problem of sediment eutrophication deals with the mobilisation of carbon in organic-rich sediments. This can include both bioaugmentation (bio-fixed microorganisms) and biostimulation (oxygen release compounds—ORC) protocols.

Waste Water Treatments

Various researchers studied waste water microbial treatment. In the wastewater industry, bacteria have traditionally been classified as Aerobic, anaerobic or facultative based on their oxygen requirements. Aerobic bacteria convert their food source (organic compounds) to energy by transferring electrons from the compounds to oxygen (electron acceptor). Anaerobic bacteria metabolises their food in the absence of oxygen and instead utilise inorganic chemicals such as iron as substitute electron acceptors and facultative bacteria can function in both aerobic and anaerobic environments.

Chemical degradation Vis-a-vis Microbial Degradation

Chemical	*Microbial*
Incomplete degradation often with dangerous residues	Mostly complete degradation with environmentally compatible residues
Energy intensive	No external energy required Economically viable for both high and low concentration residues
Not economically viable for low concentration residues	

White Rot Fungi

These fungi are very effective in cleaning up a wide range of soil pollutants like Polycyclic Aromatic Hydrocarbons, Organoachlorines, Polychlorinated Biphyenyls, Dyes, Insecticides, Fungicides, Herbicides, and Lignin.

Soil Remediation Strategies

Conversion of the soil contaminant into a biologically inert form can be accomplished by following means:

- Chemical methods: Example, addition of surfactants, acids, etc.
- Physical methods: Example, sparging, vacuum extraction, leaching, addition of cement, etc. removal of biomass and biovolatisation are two major processes involved.
- Thermal methods: Application of heat to increase extractability of the contaminants.
- Microbial remediation: The first biotreatment system targeted petroleum contaminants is shallow groundwater; now expanded to other organic pollutants as well.
- Phytoremediation: Use of plants for remediation.
- Inorganic Contaminants: Phytoremediation Strategies.

Phytoremediation

Phytoremediation is defined as use of green plants and their associated microorganisms, soil amendments and agronomic practices to remove contain or render harmless environmental contaminants. Plants have been commonly used for the bioremediation Process call Phytoremediatiion, which is to use plants to decontaminated soil by extracting heavy metals. This process is uses under satisfy environmental regulation and costs less then other alternatives. Plants that are grown in polluted soil are specialised for that process. The plants roots can extract the contaminant, heavy metals, by one of the two ways, either break the contaminant down in the soil or to suck the contaminant up, and store it in the stem and leaves of the plant. Usually the plant will be harvested and removed from the site and burned. The first plant-based remediation system was installed—300 years ago in Germany for the treatment of municipal sewage (Water Pollution).

Phytovolatisation: removal of contaminant to atmosphere; Contaminant absorbed by roots, transported through xylem and, ultimately, volatilised from the leaf surface Volatilisation of Se through plant and/or plant-microbe interaction Hg volatilisation by transgenic Arabidopsis thaliana expressing bacterial mercuric ion reductase Mercuric ion is reduced to volatile Hg(O) in the root and is volatilised from the leaf surface Phytoexctraction: exctractin of contaminants;

Phyytodegradation: Plant metabolism of contaminants, the contaminant may be metabolised by the plant enzymes in the root and/or shoot, or even in the rhizosphere (by secreted enzymes) Example: trinitrotoluene degradation by plant root nitroreductase. Other relevant plant enzymes, dehalogenase, laccase, etc.

Rhizodegradation: microbial metabolism of contaminant in rhizosphere. Rhizosphere microorganisms can play a major role in the decomposition of many organic contaminants. Example, accelerated degradation of certain pesticides, petroleum hydrocarbons, etc.

Steps involved in phytoremediation are:

- Preparation of the site for planting
- Amelioration of the physical and chemical properties of the site for plant growth
- Plants will absorb, sequester, and/or degrade the contaminant
- Agronomic practices will aim to maximise the phytoremediation effect

Useful in sandy soils with no organic matter.

Hyperaccumulator Plant Species

Plants accumulate small amounts of inorganic contaminants. The levels of target pollutants range from 0.1 to 100 mg per kg dry biomass But some plant species accumulate relatively high levels of metal contaminants (hyperaccumulators). These are taxonomically widespread in the plant kingdom.

Metal	*Plant species*	*Concentration*
Cd	Thlaspi caerulenscens	1.8 g / kg dry wt.
Cu	Ipomoea alpina	12.3 g / kg dry wt.
Co	Haumaniastrum robertii	10.2 g / kg dry wt.
Pb	Thlaspi rotundifolium	8.2 g / kg dry wt.
Mn	Macadamia neurophilla	51.8 g / kg dry wt.
Ni	Psychotria douarrei	47.5 g / kg dry wt.
	Sebertia acuminata	25% of dry sap
Zn	Thla spi. caerulenscens	51.6 g/kg dry wt.

An Example of Phytoremediation: A Patented Process

Carrots are planted on the contaminated site which can absorb and sequester DDT(dichloro-diphenyl-trichloroethane). Carrots are harvested and dried in the sun and dried biomass is incinerated to destroy the DDT. A lipid soluble contaminant moves from soil into the high-lipid content carrot roots.

Cost effectiveness is one of the greatest apparent advantages of phytoremediation, phytoremediation may cost Rs. 1-6 lakh/ha compared to Rs. 10-30 lakh/ha by other methods. Again a field of plants provides a direct, visual bioassay of the contaminant status of soil and plants are more aesthetically pleasing than any other bioremediation technique.

However there are limitations includes-The contaminant must be within the root zone or must be drawn within the root zone and the site must be large enough to make farming techniques appropriate. Phytoremediation takes much longer time than the conventional 'dig and dump' techniques It must not present health hazard or further environmental harm.

Enzymatic Bioremediation

Enzymatic bioremediation is potentially a rapid method of removing environmental pesticide residues. Applications include the treatment of residues resulting from agricultural production and processing industries, such as the treatment of irrigation waters, surface-contaminated fruit and vegetables and spent dip liquors. A specific application for some organophosphate-degrading enzymes involves detoxification of nerve agent stockpiles. Effective and affordable remediation requires highly specialised enzymes, so protein engineering techniques are being used to improve properties of various source enzymes to enhance catalytic rates, stability and substrate range. The enzymes developed for environmental remediation for specific pesticide classes also have applications as antidotes for high-dose pesticide poisonings and as prophylaxis for people at risk of high pesticide doses.

Bioremediation in Aquaculture

Microbial Process in Aquaculture

Bacteria and other microorganisms like fungi, are able to metabolise and transform numerous organic and inorganic compounds. Aerobic microbial process in aquaculture includes oxidation of organic mater to carbon dioxide and this process is the main oxygen consumer in aquaculture ponds Anaerobic Microbial Process in aquaculture include consumption of organic matter without utilisation of oxygen results in products which are not fully oxidised like alcohols and organic acids.

Microbial Mats, Biofilm and Substrate-based Aquaculture

Microbial mats are laminated heterotrophic and autotrophic vertically stratified communities typically dominated by cyanobacteria, eukaryotic microalgae like diatoms, anoxygenic phototrophic bacteria and sulphate reducing bacteria (D'amellio *et al.*, 1989). Microbial mats occur in nature as stratified communities of cyanobacteria and bacteria, but they can be cultured on large-scale and manipulated for a variety of functions. They are complex systems, but require few external inputs. Regarding aquaculture, mats were shown to produce protein, via nitrogen fixation, and were capable of supplying nutrition to tilapia *(Oreochromis niloticus).* Mats sequester or precipitate metals/radionuclides by surface absorption or by conditioning the surrounding chemical environment, thus bioconcentrating the metal/radionuclide in a small volume. Organic contaminants are degraded and may be completely mineralised. For agriculture mats hold promise as a soil amendment and nitrogen fertilizer (Judith and Phillips, 2004).

Biofilm and periphyton based complex microbial systems have been intensively studied for their beneficial effects on the productivity of carp aquaculture by independent groups working in Indian subcontinent. These investigations have shown that periphyton and biofilm based complex microbial community develops on substrate provided in the production systems and results in enhanced overall productivity from such aquaculture systems. These pioneering scientific studies involving the periphyton and biofilm based microbial interventions give rise to substrate based low-cost aquaculture especially for omnivore and herbivore freshwater fish. Systematic investigations are lacking on the role of biofertilizers in marine and brackishwater aquaculture.

Bioaugmentation and Bioremediation

Unlike in agriculture where plants utilise nutrients from the soil for primary productivity, aquatic organisms such as fish and shellfish depend on the primary producers, the plankton, for growth and development. Uptake of nutrients by plants was rendered more efficient through microbial interventions in the form of microbial biofertilizers (Nagaraju *et al.*, 2001). Studies have shown that similar nutrient utilisation also occurs naturally in aquatic ecosystems too. Of the two important nitrogen fixing bacteria, *Azotobacter* and *Azospirillum,* the former is more abundant in the freshwater aquatic ecosystems due to its aerobic nature (Das and Ayyappan, 1998). An *in vitro* evaluation of these two bacterial species revealed that the *Azotobacter* helps hi raising the PCV concentrations of the water phase and hence, phosphorous was not limiting for plankton productivity (Tripathy and Ayyappan, 2005). Similar studies on the nitrogen fixing bacteria (Garg and Bhatnagar, 1999), optimisation of exploitation of nutrients by the bacteria through manipulation of CN ratios (Jana et al 2001) were carried out that helped in achieving enhanced fresh water aquaculture production.

Bioaugmentation involves seeding of bacteria that purify aquaculture water making it less stressful to the cultured organism. The pond aquatic conditions can become toxic due to high total ammoniacal nitrogen (TAN). Under such systems, organic load in the pond water and discharge water has to be managed to avoid eutrophication. Depending upon stocking density, total phosphorus, nitrogen and suspended solids in case of a shrimp farm is reported to be as high as 321, 668 and 215,000 kg/ha/cycle, respectively (Dierberg and Kiattisimkul 1996). Mechanical treatment and effluent biofiltration systems involve a high capital cost and other technical problems making bioaugmentation a suitable alternative for aquaculture systems.

Nitrification takes place, naturally, by two groups of bacteria; *Nitrosomaonas* and *Nitrobacter*. Both being slow growing species and the practice of water exchange doesn't allow them get established in the pond and perform in an efficient manner. Thus, measures of bioaugmentation by way of providing right concentrations of bacteria on a stabilised substrate have to be implemented. Efficacy of commercial bioaugmentation products has been questionable due to the reasons of inadequate substrate, inter-specific competition, growth inhibition and insufficient acclimatisation period (Stephenson and Stephenson, 1992). Hence, considerable research in isolating and establishing native beneficial bacteria on suitable substrates has to be carried out.

Bagass-assisted bioremediation of ammonia from shrimp farm wastewater was studies by Krishnani *et al* (2006). They indicated that the bagass-assisted substrate supported by the rich bacterial biofilm could remove TAN to an extent of 95 per cent by 144 h. Removal of TAN using enriched immobilised cultures of autochthonous nitrifying bacteria has several potential advantages: longer biomass retention in pond water, maintenance of high microbial cell density, optimisation of microbial growth and metabolic rates and protection from inhibitory compounds (Travieso et al 1995; Yang 1997).

Bioaugmentation and bioremediation measures and technologies are being developed in various institutes and laboratories in India. The NCAAH (National Centre for Aquatic Animal Health), Cochin; Marine Biotechnology (MIRCEN) lab headed by Dr. Karunasagar at College of Fisheries, Mangalore; Central Institute of Brackishwater Aquaculture (CIBA), Chennai are notable among them. Several non-pathogenic bacteria are known for their anti-pathogenic effects and hence, have tremendous potentials of application in managing the microbial balance in favour of the host and in reducing the pathogen load. Lactic acid bacteria (Panigrahi *et al.*, 2004,2005; Gatesoupe, 1994; Ajitha *et al.* 2004), *Caronbacterium* (Robertson *et al.* 2000), *Bacillus* (Gatesoupe, 1991; Rengpipat *et al.* 1998; Vaseeharan and Ramasamy, 2003), *Vibrio* (Austin et al 1995), *Pseudomonas* (Chythanya *et al.* 2002; Vaseeharan *et al.* 2003; Vijayan *et al.* 2005), etc., have been some of the bacterial species known for their pathogen antagonism.

A bioreactor technology (patent yet to be granted) meant for nitrifying seawater before and after larval rearing was developed by NCAAH. The team also claims to have developed a cyanobacterial system consisting of two species of genera *Synechocistis* for the control of vibrios in aquaculture ponds. Work on the isolation and characterisation of bacterial antagonists has been carried out in India by several active working groups.

Genomic Approaches

Researchers now characterise the microorganisms that are important in bioremediation and can evaluate their physiology using a combination of genome enabled experimental and modelling techniques. New environmental genomic techniques coupled with metabolomics and proteomics offer various possibility for studies and innovations. Microbial removal of contaminants from the environment often takes place without human intervention. This has been term intrinsic bioremediation.

Genetically Engineered Microorganisms

There are instances where natural populations are not suitable for use in the remediation of polluted sites and therefore the utilisation of genetically engineered microorganisms (OEMs) is being considered for in-situ bioremediation of contaminated ecosystems. Remediation continues to be improved, with use of microorganisms (both genetically modified organisms (GMOs) and naturally occurring ones, including extremophiles) to clean contaminated areas (David *et al.*, 1995; Tebo, 1995). Off shore or other contamination of marine ecosystems resulting from toxic organics may be amenable to *in situ* remediation by naturally occurring consortia of microorganisms in sediments (Deming, 1998).

Dewitt (2000) mentioned about the *Deinococcus radiodurans*—the most radiation resistant organism yet known which was engineered to express an *Escherichia coli* enzyme that converts ionic mercury to a less toxic form and a pseudomonas enzyme that breaks down toluene. Again the isolation of novel bacterium that destroys vinyl chloride, the breakdown product of chlorinated hydrocarbon as a part of its energy metabolism raises the prospect of effective bioremediation for contaminated water reservoir (Jianzhong *et al.*, 2003).

What Microorganisms are here?

A significant advance in the field of microbial ecology was the finding that the sequences of highly conserved genes are found in all microorganisms and the 16S rRNA approach is very handy The 16S rRNA genes, could provide a phylogenetic characterisation of the microorganisms that comprise microbial communities (Pace *et al.*, 1986; Amann *et al.*, 1995). By analysing

16s rRNA sequences in contaminated environments, the phylogenetic placement of the microorganism that are associated with bioremediation processes can be detected (Rogers, S. L. & McClure, 2003; Amann *et al.*,1995). However, the knowledge of the phylogeny of the organisms associated with bioremediation dose not necessarily predict important aspects of their physiology (Snoeyenbos *et al.*, 1997, Pace *et al.*,2000). For example, microorganisms with 16S rRNA sequences closely related to the TCE degrader D ethanogenes can differ hi the chlorinated compounds that they can degrade (He *et al.*, 2003; Bunge *et al.*, 2003).

What can be the Functions of Microorganisms?

Analysis of genes involved in bioremediation: Examining the presence and expression of the key genes involved in bioremediation can yield more information on microbial processes (Rogers and McClure, 2003). In general there is a positive correlation between the relative abundance of the genes involved in bioremediation and the potential for contaminant degradation 18.36 (Rogers, and McClure, 2003; Schneegurt and Kulpa 1998).

Molecular Mechanisms Involved?

The genes for bioremediation can be present but not expressed. Therefore, there has been an increased emphasis on quantifying the levels of mRNA for key bioremediation genes. Increased mRNA concentrations can be, at least qualitatively, associated with higher rates of contaminant degradation (Schneegurt and Kulpa 1998). For example the reduction of soluble ionic murcury, Hg(II), to volatile Hg(o), is one mechanism for removing mercury from water; the concentration of mRNA for merA—a gene involved in Hg (II) reduction—was highest in mercury contaminated waters with the highest rates of Hg (II) reduction (Nazaret *et al.*, 1994).

In a series of experiments (Panigrahi *et al.*, 2004 a & b, 2005 and 20066) attempt was made to draw some relations between the different characteristics of probionts and the immune response they elicit in aquatic animals. The ability of three different bacterial species to induce non-specific immune responses and the expression of certain cytokine genes interleukins (IL- pi and 2), Tumour necrosis factor (TNF-a 1 and 2) and transforming growth factor (TGF-P) were evaluated.

Whole Genome Sequencing

Whole genome sequencing is especially helpful in promoting the understanding of bioremediation relevant microorganisms, whose physiology has not previously been studied in detail. For example, as noted earlier, molecular analyses have indicated that Geobacter species are important in the bioremediation of organic and metal contaminants in subsurface environments. Studies revealed that *Geobacter metallireducens* specifically produces flagella only when the organism is growing on insoluble Fe(III) or Mn(IV) oxides.

With the completed genome sequences, it is possible—using whole-genome DNA micro arrays to analyse the expression of all the genes in each genome under various environmental conditions. Using proteomic techniques, it is possible to identify which proteins are expressed (Mohn and Tiedje, 1992).

Identifying regulatory circuits in these organisms (National Academy Press, Washington DC, USA, 1993) through genome-wide expression analysis is significant as the mechanisms that control the regulation of the catabolic and respiratory genes are the most important in bioremediation.

Metabolomics: The Metabolic Potential

Evaluation of the metabolic state of bioremediating microorganisms through analysis of the mRNA concentrations for key genes that are involved in responding to Stress could help to identify- modifications to contaminated environments that might promote bioremediation. Global insight into the metabolic potential and activity of microorganisms living in contaminated environments.

It is necessary to have a more holistic view of metabolism in models that can describe the outcome of the thousands of individual reactions that are simultaneously taking place in a microbial cell.

As predicted from the annotated genome the fluxes of metabolic inputs and outputs of all these reactions to describe the overall pathway and output of cell metabolism under a given set of environmental conditions (Covert *et al.* 2001). The transcriptional regulation of gene expression is understood from experimental data, this can be added as an additional constraint (Covert and Palsson 2002).

Genomic DNA extracted from environmental samples can be cloned into large (-40-100 kb) cloning vectors, such as Bacterial artificial chromosome (BACs), to make a library of environmental genomic DNA. (DeLong 2002; Handelsman and Goodman 2002). A global analysis of which genes are being expressed under various conditions in contaminated environments wall reveal the metabolic status of the microorganisms and indicate environmental modifications, possibly as simple as the addition of a trace nutrient, which might accelerate the bioremediation.

Functional Genomics

Expression data could be incorporated into in silico-models that can predict not only the activity of the individual microorganisms responsible for important bioremediation reactions, but also the interactions of these organisms with other microorganisms in community, such studies should yield a basic understanding of the diversity, function, evolution and ecological interactions of microbial communities. (DeLong 2002; Rodriguez-Valera 2002), and could help answer the age-old question in microbiology of what defining a microbe (Rodriguez-Valera 2002). Suicidal GEM.

Again some of the ethical questions are being raised for the fate of the genetically engineered microbes (GEM). The deliberate release of OEMs into the environment is a subject of considerable public concern. The potential risks associated with the release of GEMs into the environment has led to the construction of biological containment systems by which bacteria are killed in a controlled suicide process. It consists of two different components, a killing element designed to induce cell death and a control element which modulates the expression of the killing function.

Conclusions

The demand for green technologies and sustainability is increasing with food and biosecurity. Now, when agri/ aqua culturist is moving towards organics and has to follow the effluent discharge guidelines as stipulated by the Environmental Protection Agencies (EPA), bioremediation or bioaugmentation can be a great aid. The substrate, based aquaculture, the concept of biofilm, biocapsules, probiotics and microbial mats are being increasingly recognised in commercial aquaculture.

The application of genome-enabled techniques to the study of bioremediation is clearly in its infancy. There are many technical issues that needs to be addressed before some of the more novel approaches, such as environmental genomic sequencing and arrays, or in silico modelling of multi-species consortia can be routinely employed in this field, also the nanotechnology brings lots of promise in this regards. However, for bioremediation to advance as a science, there is strong need for the comprehensive understanding of physiological properties that genome-enabled approaches can provide.

The ultimate goal of such studies is the coupling of models of microbial growth and metabolism in contaminated environments with existing geochemical and hydrological cycle to predict accurately the microbial assisted natural attenuation of contaminates or the likely outcome of engineered strategies to accelerate bioremediation. In view of the genomic approach and the vast potential lies therein, we can attribute the future coming to be the 'genomics era' of bioremediation. Truly microorganisms are nature's original recycle agents.

Bioremediation of Environmental Degradation

A rapid pace of industrialisation coupled with uncontrolled exploitation of nature has resulted in continuous dumping of industrial by-products, hazardous chemicals, nuclear wastes, deforestation in vast areas of earth and pollution of river basins, lakes and seas. In his quest for wealth and comfort, man has distributed the delicate ecological balance between the living and non-living components of the biosphere. The unfavourable conditions created by man not only threatened the survival of man himself but also other living organisms. The number of plant and animal species likely to become rare, threatened, endangered or near extinction in the Red Data Book of IUCN is increasing with time.

A pollution free environment is essential for sustenance of life on earth. World Health Organisation (WHO)'s call "Health for all by 2000 AD" gives prominence to all round improvement of environment for better health and preventing diseases. Therefore, massive programmes for improving the already deteriorated environment are under way in most countries especially in developing countries of Asia, Africa and Latin America. Among various methods, the "Biological Approach" to control the environmental pollution has been demonstrated to be more useful and effective in the recent years.

Control of Pesticide and Chemical Fertilizer Pollution

Intensive agriculture with the use of chemical fertilizer and pesticide has no doubt, resulted in manifold increase in the crop productivity. However the adverse effects of these chemicals are clearly visible on soil structure, soil microflora, surface and ground water and air. The amount of pesticides applied worldwide is estimated to be 2.5 million tones. Of this, 50-60 per cent herbicides, 20-30 per cent insecticides and 10-20 per cent are fungicides. World Health Organisation (WHO) has estimated that globally 3 million people are poisoned by these pesticides and weedicides every year of which 20,000 die. A majority of the pesticide poisoning have been reported from developing countries which use only 25 per cent of the global pesticides and chemical fertilizers.

Potential alternatives to chemical fertilizers and pesticides are available in nature. Such biological and natural inputs include plant origin pesticides, biopesticides, bioherbicides, biofertilizers and organic manures. A committee of the National Academy of Science, USA headed by Dr. Donald Kennedy of Stanford University have suggested that the biological agents should be used to control the pests and weeds. The biological agents are inexpensive, pose no risk to human health, have a long term capability to control pests and can improve the soil fertility. Moreover, they help us to avoid the dumping of thousands of tones of agrochemicals on the earth every year and give us a safe environment (Jogadan, 1995, Singh, 1996).

Origin Pesticides

The formulations of many plant products such as Neem *(Azadirachta indica)*, Pongamia *(Pongamia glabra)*, Tulsi *(Ocimum sanctum)*, Sorghum, Marigold and Bouganvillea have been reported to be effective against number of diseases and pests. Appropriate substitution of chemical pesticides with these plant products could certainly be more eco-friendly (Sharma, 2001, Singh, 1996).

Biofertilizers

These are microorganisms (bacteria, fungi, algae) which are used to fix the atmospheric nitrogen to solubulise and make nutrients available to plants to synthesise other growth promoting substances. Important bacterial fertilizers include Azospirilum, Rhodosprilum, Azotobacter, Pseudomonas, Klebsilla, Rhizobium and Clostridium. The Mycrrhizal biofertilizers including VA Mycorrhiza, Aspergillus awamori. A niger and Penicillium digitatum help in phosphorous fixation in soil and improve water transport. Algae that are used as fertilizers include Anabaena, Nostoc, Aulosira, Tolypothrix, Gloeotrichia and Scytonema. Azolia is a tiny, floting fresh water fern, which is used as biofertilizer. It has a symbiont blue green alga (Anabena azollae) which helps in nitrogen fixation. Azolia has a capacity to fix upto 90 Kg N/ha in soil. Other sources of biofertilizers are green manures like sesbania and sunhemp, etc. (S.J Ignacimuthus, 1998, Kumar *et. al.* 1999).

Biopesticides

Biopesticide agents are abundantly available in nature. Several pathogens including viruses such as nuclear polyhedrosis virus (NPV) granulosis virus (GV), bacterial like *Bacillus thuringenesis,* fungi like *Metarhizium,* protozoa like *Schizogregrine* causes diseases in insects and destroy them. Insect-parasitoids (parasites thriving on insects) like *Chelonus, Trichoderma, Eriborus, Elamus, Bracon and Trichospilus* attacks the insects pest and destroy them. Insect predators such as *Chloconus Cryptolaemus, Crytorhinus* and predatory, *Neoplectana and Heterorhabdtis* can also be used for pest control (Ignacimuthus, 1998).

Bioherbicides and Bioweedicides

Weeds are undesirable plants growing in the field competing with the crop plants for nutrients, moisture and space. Biological agents like insects, fungi and bacteria are reported to be effective against a number of weeds. The insects like weevil *(Neochitinia biocolorata)* is effective against congress grass*(Parthenium histerophorus).* Formulations of soil borne fungus, *Phytophthora palmivora* is effective against strangler vine and of *Collectrotrichum gloesporoides* against northern jointvetch in rice and soyabean. These two mycoherbicides have been registered in USA for commercial use.

Application of Organic Manure

Addition of organic manure like farmyard manure, crop residues, biogas slurry, crop wastes, earth worms and composts change the rhizosphere environment by effecting porosity, aeration, temperature, water holding capacity and root microflora. Role of earthworms is also important as they play a critical role in aeration, soil turnover and promotion of microflora required for plant growth. Addition of organic amends to soil have also been proved effective in controlling number of diseases and nematodes (Sharma *et al.* 2001).

Transgenic Plants

Transgenic plants broadly speaking, refers to those plants in which functional foreign genes have been inserted in their genomes. Anumber of methods are now available to produce transgenic plants which include Agrobacterium mediated DNA up take, electroporation, microinjection or microprojectile bombardment. Transgenic plants resistant to herbicide insects and viruses have already been produced. Initially the production of transgenic plants was restricted to dicotyledons, however it has now been extended to a number of monocotyledonous plants like rice, wheat and maiza. These plants under field conditions have also maintained increased level of insects and herbicide resistance and higher levels of virus protection in near future, these genetically engineered plants will significantly reduce the application of chemical fertilizers and pesticides in the field (Gupta, 1998).

Biocontrol of Gaseous Pollutants

The atmosphere is an insulting blanket around the earth and is the source of all essential gases. Without atmosphere there will be no wind, no cloud, no lightining and no rain. It also acts as a shield against lethal UV radiation. However, it is getting polluted day-by-day by the poisonous gases emitted from the industries, thermal power stations, automobiles and domestic combustion. The principal gaseous pollutants emitted from different sources are carbondioxide (CO_2), carbonmonoxide (Co), sulphurdioxide (SO_2), hydrogen sulphide (H_2S), nitrous oxide (N_2O), nitrogenoxides (NO_2) and chlorofluoro carbons (CFC).

Plantation of pollutants absorbing plants: There are many plant species which have been identified to have the ability to absorb the gaseous pollutants. These plants include; *Pongamia pinnata, Dalbergia sissoo, Bambusa vulgaris, Caesalpinia pulcherima, Azadiyachta indica, Tectona grandits, Therminalia arjuna, Mimuspos elingi, Cassia siamea, Cassia fistula, Acacia nilotica, Tamarindus indica, Ficus benghalensis and Mangifera indica.* Creation of thick vegetation cover by large-scale plantation of these tree species near the industrial areas, power plants and along the streets and highways will markedly reduce the poisonous effect of these chemicals (Patra, 1999, Sharma, 2001).

Microb-based Air Purification

Gaseous pollutants emitted from the industries are usually removed by using electrostatic precipitators or other electronic devices. But now several "microbes" having the capability to utilise pollutants as nutrients have been identified several air-purification plants have been developed in which the culture of these microbes are used to clean the air. A survey carried out by Institute of Hygiene and Physiology at Zurich, Switzerland, showed that about 40 such microbe-based air purification plants are there in Japan. These microbe-based plants are of two types: (i) bioscrubber, (ii) biofilters. In bioscrubber the polluted gas is cleaned by spraying it with a scrubbing liquid which consists of microbes and their nutrients. The biofilters consist of leather, compost and peat. The filter bed should always be kept moistened for the survival of the microbes. The volatile substances and other pollutants are removed when the polluted gas is allowed to pass through the filter (Sharma *et al.*, 2001, Pelczar *et al.*, 2000, Jogadan, 1995).

Bacteria for CFC Degradation

The scientists of the United States Technological Survey have discovered certain bacterial strain that can break down one of the most hazardous pollutants "chlorofluro carbons". These bacterial strains are found along the Potame river in Virginia in pond, marsh and swampy sediments of South Carolina (Sharma *et al.*, 2001).

Sulphur Compound Degrading Bacteria

The sulphur compound, SO_2 and H_2S are extremely harmful gaseous pollutants which are responsible for acid rain. These bacteria inhabiting the hot water springs have the ability to degrade sulphur compounds. The coal is another source of sulphur pollutants. The sulphur degrading bacteria can be used to produce sulphur free coal. A small scale plant in Okie, Japan has successfully used the bacteria to produce sulphur free coal.

Biocontrol of Sewage and Industrial Effluent Pollution

Sewage is the water—borne waste derived from houses and animals which include human excreta, paper, cloth pieces, soap, detergents, etc. Due to the accumulation of sewage in the water bodies, the water become unfit for drinking and other domestic uses. The industrial effluents include both organic and inorganic pollutants such as oil, greases, phenols, toxins, salts, dyes, heavy metals (chromium, lead, cadmium and mercury), suspended solids, plastic and other metallic wastes (S.J. Ignacimuthus, 1998).

Use of Water Hyacinth for Wastewater Treatment

Recently United States National Aeronautics and Space Administration (NASA, USA) has recommended the use of water hyacinth *(Eichhornia crassipes)* and dusk weeds (Spirodeia, Lemna sp. Wolffia ssp.) for treatment of wastewater. These plant species have the ability to absorb and accumulate the heavy metals and other elements such as sulphur, iron, potassium, etc. Useful minerals can be extracted from these plants by harvesting them periodically. The harvested weeds after metal extraction can be used for biogas production and also as fertilizer. Water hyacinth can also absorb various organic compounds like phenois, benzene and toxophene, etc. and certain radio active toxicants which are very different to remove from wastewater.

Earthworms as Sewage Cleaners

The scientists of the Bhawalkar Earthworm Research Institute at Pune have recently developed a method that makes use of the earthworms to remove pollutants from water. In this method the wastewater is filtered through "vermin-filter" formed by earthworms and worm-casts. The earthworms convert the impurities into worm-casts and the resulting clean water can be used for gardening. The worm-cast which harbours a wide spectrum beneficial microflora and organic nutrients, can be periodically harvested and used as fertilizer (Sharma *et al.* 2001).

Mushrooms for Sewage Treatment

The most recent method developed to control the heavy metals and radio active waste pollution is "biosorption" (Absorption of pollutants by biological agents). The mushrooms like *Ganoderma lucidium and Glucidium* have been identified for treatment of effluents from metal planting, rare earth metal processing industries and waste from nuclear power plants. The biggest advantage of using the "biosorbents" is that the metals can be extracted from them by simple techniques and then can be reused. (Sharma *et al.* 2001).

Biogas Production

In many countries, methonogenic bacteria are employed in sewage treatment. The methonogenic bacteria are able to utilise acetate, methanol and other organic waste acetate for the production of methane. The methane gas thus produced is collected and used as fuel (Sharma *et al.*, 2001, Dey, 1995).

Biological Control of Noise Pollution

Noise is defined as the unwanted or unpleasant sound that causes discomfort. It not only causes irritation or annoyance but it also constricts the arteries, increases the flow of adrenaline, forces the heart to work faster and reduces the working efficiency. It also causes auditory fatigue and deafness.

There are many plant species which have been reported to have the ability to absorb and dissipate sound energy and act as a "buffer zone". These plants species include *Azadimchta indica, Bauhinia purpurea, Ficus benghalensis, Cassia fistula, Langeroistroemia reginae, Saraka ashoka, Salmalia malabarica and*

Erythryna variegate. Large-scale plantation of these species along the road sides, highways and around the airport will help in attenuating noise and purifying air (S.J. Ignacimuthus, 1998, Sharma *et al.*, 2001, Patra, 1999).

Biocontrol of Spilled Oil and Deposits

Oil spills from tankers on land surface and on seas have been recognised as a major environmental hazard. This spilled oil is believed to not only destroy of habits of aquatic animal but also create health problem for local residents. The chemical dispersants used to clear the oil spills are believed to care environmental problem due their toxic nature and persistence in environment. In recent years genetic engineered oil utilising bacteria have been produced which grow rapidly on oil. A strain of *Pseudomonas aerginosis* has been developed by Ananda Chakraborthy at University of Illinois, Chicago (USA), which produces a glycolipid emulsifier that reduces the surface tension of oil water interface and helps in removal of oil from water. The microbial emulsifier is non-toxic and biodegradable and thus being ecologically sound (Jogadan, 1995, Gupta, 1996).

Bio-sensors

It is a biophysical device, which will detect the presence and measure the quantities of specific substances in a variety of environments.

A bio-sensor equipment consists of:

i. A biological sensing agent.

ii. A device for collecting the product obtained from the interaction between the substrate and the bio-sensor.

iii. A device for measuring the quantity of these products, thus indirectly giving an estimate of the substrate.

iv. Used in medicine as well as industry for monitoring environment.

For instance enzymatic tests have been designed to detect pesticide contamination in water and cost only 3.00$ (US)/ test compated to 150-300$ (US) for a laboratory test. Thus bio-sensors can detect pollution problem at an early stage, so that environmental monitoring is possible.

Biodegradation of Xenobiotics and Toxic Wastes

Bacteria can be modified genetically to develop catabolic pathways for degradation of Xenobiotics (Wastes from non-biological systems) and other waste materials. Some of the genetically engineered bacterial strains for biodegradation of waste material are *Pseudomonas capalia* degrading 2,4,5 - trichloro-phenoxyacetir acid, *P. Putida* and other Sp. Degraded 2,2,5 dischloroaromatics, *Alcalogences sp.* degrades dichlorophenoryaletic acid, mixed chlorophenolis, 1,4-dhlorobenzene and Acinetoacter sp. Degrades 4-chlorobenzent, etc. (Jogadan, 1995).

Bioplastics

Plastic (Polymers) can be degraded by three different processes either independently or, in combination with: (i) light or, high energy radiation, (ii) heat, (iii) microbes. In response to increasing public concern over environmental hazard caused by plastic, many countries are conducting various solid waste management programmes including plastic waste reduction by development of biodegradable plastic material. Some biodegradable plastic materials under development are: (i) poly-hydroxy alkanoates (PHAs), (ii) poly-lctides, (iii) aliphatic polyesters (iv) polyusaccharides and (v) Co-polymers and/or bleds (vi) poly/betahydroxy butyrate (PHB) of above. It is possible to produce biodegradable plastics from bio-wastes with the help of a syntrophic population of microbes of diverse origin (Jogadan, 1995, Kalia, 2000, Singh, 1996).

Biofilms

A biofilm can be defined as matrix-enclosed bacteria populations adhering to one another and/or to surface or interfaces. Biofilms are found in virtually all environments that carry sufficient nutrients. They are many times more resistant to anti-bacterial agents (Kumar & Anand, 1998). Biocontrol of biofilm formation in the adsorption of bioactive compounds onto a proteinaceous anti-microbial compounds that exhibits bactericidal properties. Nisin is an anti-microbial peptide that is an effective food pathogen inhibitor. It is used for the control of Clostoridium botulinum spores in cheese spread. Other biocontrol are lactic cultures and enzumes (Kumar & Anand, 1998).

Strains of Lactococcus lactus is a wide range of bacteriocins, of which the most important is nisin which is active against Gram-positive bacteria (Stiles & Holzapfel, 1997). The use of enzymes to remove and destroy bacterial biofilms has been considered (Johanson, Falholt & Gram, 1997). Biofilm applications also include the degradation of certain pollutants. In these applications, new genetic information has been introduced into the microorganisms that form the biofilms. This process is called bioaugmentation. In this way, new biodegradation pathways fro metabolic conversion of the pollutants are introduced (Jackson *et al.*, 1996).

Conjugational transfer of a plasmid in a biofilm engaged in the degradation of benzyl alcohol (Christensen *et al.*, 1998). Another way of "remodelling" a bacterium strain by selecting based on the bacteria's behaviour, such as adhesion to chosen substrate (Vidal *et al.* 1998). There are biofilms for bioremediation of contaminated soil and water, biofilms for wastewater treatment and vapour phase biofilters for the prevention of industrial air pollution (Schreuders *et al.* 2002)

5

Role of Biofilm

A biofilm is a complex aggregation of microorganisms marked by the excretion of a protective and adhesive matrix. Biofilms are also often characterised by surface attachment, structural heterogeneity, genetic diversity, complex community interactions, and an extracellular matrix of polymeric substances.

Single-celled organisms generally exhibit two distinct modes of behaviour. The first is the familiar free floating, or planktonic form in which single cells float or swim independently in some liquid medium. The second is an attached state in which cells are closely packed and firmly attached to each other and usually a solid surface. The change in behaviour is triggered by many factors, including quorum sensing, as well as other mechanisms that vary between species. When a cell switches modes, it undergoes a phenotypic shift in behaviour in which large suites of genes are up- and down-regulated.

We may not be familiar with the term "biofilm," but we have certainly encountered biofilm on a regular basis. The plaque that forms on your teeth and causes tooth decay is a type of bacterial biofilm. The "gunk" that clogs the drains is also biofilm.

Biofilm forms when bacteria adhere to surfaces in aqueous environments and begin to excrete a slimy, glue-like substance that can anchor them to all kinds of material—such as metals, plastics, soil particles, medical implant materials, and tissue. A biofilm can be formed by a single bacterial species, but more often biofilms consist of many species of bacteria, as well as fungi, algae, protozoa, debris and corrosion products.

Essentially, biofilm may form on any surface exposed to bacteria and some amount of water. Once anchored to a surface, biofilm microorganisms carry out a variety of detrimental or beneficial reactions (by human standards), depending on the surrounding environmental conditions.

Formation

Formation of a biofilm begins with the attachment of free-floating microorganisms to a surface. These first colonists adhere to the surface initially through weak, reversible van der Waals forces. If the colonists are not immediately separated from the surface, they can anchor themselves more permanently using cell adhesion molecules such as pili.

The first colonists facilitate the arrival of other cells by providing more diverse adhesion sites and beginning to build the matrix that holds the biofilm together. Some species are not able to attach to a surface on their own but are often able to anchor themselves to the matrix or directly to earlier colonists. It is during this colonisation that the cells are able to communicate via quorum sensing.

Once colonisation has begun, the biofilm grows through a combination of cell division and recruitment. The final stage of biofilm formation is known as development, and is the stage in which the biofilm is established and may only change in shape and size. This development of biofilm allows for the cells to become more antibiotic resistant.

Properties

Biofilms are usually found on solid substrates submerged in or exposed to some aqueous solution, although they can form as floating mats on liquid surfaces. Given sufficient resources for growth, a biofilm will quickly grow to be macroscopic. Biofilms can contain many different types of microorganism, e.g. bacteria, archaea, protozoa and algae; each group performing specialised metabolic functions. However, some organisms will form monospecies films under certain conditions.

Extracellular Matrix

The biofilm is held together and protected by a matrix of excreted polymeric compounds called EPS. EPS is an abbreviation for either extracellular polymeric substance or exopolysaccharide. This matrix protects the cells within it and facilitates communication among them through biochemical signals. Some biofilms have been found to contain water channels that help distribute nutrients and signalling molecules. This matrix is strong enough that under certain conditions, biofilms can become fossilised.

Bacteria living in a biofilm usually have significantly different properties from free-floating bacteria of the same species, as the dense and protected environment of the film allows them to cooperate and interact in various ways. One benefit of this environment is increased resistance to detergents and antibiotics, as the dense extracellular matrix and the outer layer of cells protect the interior of the community. In some cases antibiotic resistance can be increased 1000 fold.

Persister Cells

Kim Lewis of Northeastern University has discovered that a small fraction of cells within *E. coli* biofilms are dormant within the biofilm and almost immune to the effects of antibotics because of their very low level of metabolic activity. Once antibiotic levels drop, these "persister cells" become active and repopulate/reconstitute the biofilm. Persisters are not mutants, but phenotypic variants of the wild type. He developed a method to isolate persisters and obtained their gene profile, which pointed to a dormancy programme that is turned on in these cells. He found that proteins known as "toxins" forming toxin/antitoxin modules are largely responsible for persister formation. "Toxins" appear to be the exact opposite of what their name suggests, they reversibly block important processes such as translation, protecting the cell from bactericidal antibiotics that are ineffective against inactive targets.

This finding provides a general solution for the long-standing riddle of biofilm tolerance. His current focus is to understand the workings of this dormancy programme. The same programme may be responsible for other puzzling phenomena in microbiology, such as latent forms of infectious pathogens, and uncultivability of most bacterial species on synthetic media. He is also searching for drugs that can kill persister cells.

Examples

Biofilms are ubiquitous. Nearly every species of microorganism, not only bacteria and archaea, have mechanisms by which they can adhere to surfaces and to each other.

Biofilms can be found on rocks and pebbles at the bottom of most streams or rivers and often form on the surface of stagnant pools of water. Biofilms are important components of foodchains in rivers and streams and are grazed by the aquatic invertebrates upon which many fish feed.

Biofilms grow in hot, acidic pools in Yellowstone National Park (USA) and on glaciers in Antarctica.

In industrial environments, biofilms can develop on the interiors of pipes, which can lead to clogging and corrosion. Biofilms on floors and counters can make sanitation difficult in food preparation areas.

Biofilms can also be harnessed for constructive purposes.

For example, many sewage treatment plants include a treatment stage in which waste water passes over biofilms grown on filters, which extract and digest organic compounds. In such biofilms, bacteria are mainly responsible for removal of organic matter (BOD); whilst protozoa and rotifers are mainly responsible for removal of suspended solids (SS), including pathogens and other microorganisms. Slow sand filters rely on biofilm development in the same way to filter surface water from lake, spring or river sources for drinking purposes.

Biofilms are also present on the teeth of most animals as dental plaque, where they may become responsible for tooth decay.

Biofilms and Infectious Diseases

Biofilms have been found to be involved in a wide variety of microbial infections in the body, by one estimate 80 per cent of all infections. Infectious processes in which biofilms have been implicated include common problems such as urinary tract infections, catheter infections, middle-ear infections, formation of dental plaque, gingivitis, coating contact lenses, and less common but more lethal processes such as endocarditis, infections in cystic fibrosis, and infections of permanent indwelling devices such as joint prostheses and heart valves.

Biofilms are present on the removed tissue of 80 per cent of patients undergoing surgery for chronic sinusitis. The patients with biofilms were shown to have been denuded of cilia and goblet cells, unlike the controls without biofilms who had normal cilia and goblet cell morphology. The species of bacteria from interoperative cultures did not correspond to the bacteria species in the biofilm on the respective patient's tissue. In other words, the cultures were negative though the bacteria were present.

Bradley Smith of the University of Notre Dame reports on a new staining technique to differentiate bacterial cells growing in living animals from allery-inflamed tissue.

Industrial Significance

Microbial biofilms on surfaces cost the nation billions of dollars yearly in equipment damage, product contamination, energy losses and medical infections. Conventional methods of killing bacteria (such as antibiotics, and disinfection) are often ineffective with biofilm bacteria. The huge doses of antimicrobials required to rid systems of biofilm bacteria are environmentally undesirable (and perhaps not allowed by environmental regulations) and medically impractical (since what it would take to kill the biofilm bacteria would also kill the patient!). So new strategies based on a better understanding of how bacteria attach, grow and detach are urgently needed by many industries. Conversely, microbial processes at surfaces also offer opportunities for positive industrial and environmental effects, such as bioremediating hazardous waste sites, biofiltering industrial water, and forming biobarriers to protect soil and groundwater from contamination.

Key to Understanding and Controlling Bacterial Growth in Automated Drinking Water Systems

Biofilms are a collection of microorganisms surrounded by the slime they secrete, attached to either an inert or living surface. You are already familiar with some biofilms: the plaque on your teeth, the slippery slime on river stones, and the gel-like film on the inside of a vase which held flowers for a week. Biofilm exists wherever surfaces contact water.

As in any water system, 99 per cent of the bacteria in an automated watering system is likely to be in biofilms attached to internal surfaces. Biofilms are the source of much of the freefloating bacteria in drinking water, some of which can cause infection and disease in laboratory animals.

One common biofilm bacteria, *Pseudomonas aeruginosa,* is an opportunistic pathogen which can infect animals that have suppressed immune systems. Besides being a reservoir of bacteria which can affect animal health, biofilms can also cause corrosion in stainless steel piping systems.

Understanding bacteria in biofilms is one step towards preparing for the future. Animal facilities can meet the most demanding water quality requirements by supplying chlorinated reverse osmosis water and by maintaining water quality through flushing and sanitisation. But what if chlorine use in animal drinking water is prohibited? Or what if water quality requirements become even more stringent with the use of new specialised animals? In order to design and operate automated watering systems that deliver the cleanest water possible we should understand how biofilms develop, some of the problems they can cause, and how biofilm growth can be controlled.

Early biofilm formation and the effects of anti-microbial agents on orthodontic bonding materials in a parallel plate flow chamber Decalcification is a commonly recognised complication of orthodontic treatment with fixed appliances. A technology, based on a parallel plate flow chamber, was developed to investigate early biofilm formation of a strain of *Streptococcus sanguis* on the surface of four orthodontic bonding materials: glass ionomer cement (Ketac Cem), resin-modified glass ionomer cement (Fuji Ortho LC), chemically-cured composite resin (Concise) and light-cured composite resin (Transbond XT). *S. sanguis* was used as it is one of the primary colonisers of dental hard surfaces. Artificial saliva was supplied as a source of nutrients for the biofilms. The effects of two commercially available mouthrinses (i.e. a fluoride containing rinse and chlorhexidine) were evaluated. Initial colonisation of the bacterium was assessed after 6 hours of growth by the percentage surface coverage (PSC) of the biofilm on the disc surfaces.

There were statistically significant differences in bacterial accumulation between different bonding materials ($P < 0.05$), Concise being the least colonised and Transbond XT being the most colonised by *S. sanguis* biofilms. All materials pre-treated with 0.05 per cent sodium fluoride mouthrinse showed more than 50 per cent reduction in biofilm formation. The 0.2 per cent chlorhexidine gluconate mouthrinse caused significant reduction of biofilm formation on all materials except Ketac Cem.

This *in vitro* study showed that the use of a chemically-cured composite resin (Concise) reduced early *S. sanguis* biofilm formation. Also, fluoride had a greater effect in reducing the PSC by *S. sanguis* biofilms than chlorhexidine. Rinsing with 0.05 per cent sodium fluoride prior to placement of orthodontic appliances is effective in reducing early biofilm formation.

Dental plaque is present in both healthy and diseased individuals and is the aetiological agent of the two most prevalent diseases affecting industrialised societies: dental caries and periodontal disease (Theilade, 1989). Dental plaque is an example of a biofilm which is defined as a community of bacteria and their extracellular polymers that are attached to a surface (Wilson, 2001).

The build-up of dental plaque on teeth and dental materials is a problem in orthodontics (Ogaard *et al.*, 1988a). Despite advances in orthodontic materials and techniques in recent years, the development of enamel decalcification and dental caries around orthodontic appliances continues to be a problem. Nearly 50 per cent of orthodontic patients exhibit clinically visible white spot lesions during treatment that last approximately 2 years (Basdra *et al.*, 1996). Because it is known that white spot lesions form early (O'Reilly and Featherstone, 1987), it is important to understand the effects of biofilm build-up around orthodontic bonding materials following placement of fixed orthodontic appliances.

Composite resins and glass ionomer cements (GICs) are widely used in securing fixed orthodontic devices. During the 1990s, a major development of hybridisation technology underlying composite resins and GICs popularised the use of resin-modified GICs (RMGICs) and polyacid-modified composites (compomers) in orthodontics.

Composites used as orthodontic direct bonding adhesives

have a polymeric matrix that can host and nurture a variety of aerobic and anaerobic microorganisms acting alone or in combination (consortia). Their accumulation can lead to premature debonding and eventually enamel decalcification and periodontal diseases (Matasa, 1995). The use of fluoride-containing orthodontic bonding materials both *in vitro* (Fox, 1990; Badawi *et al.*, 2003) and *in vivo* (Banks *et al.*, 1997) has shown variable results in helping to prevent or reduce decalcification. Significantly elevated levels of fluoride in plaque have been found adjacent to GIC-retained brackets compared with those adjacent to composite resin-retained brackets up to 6 months after the onset of treatment (Hallgren *et al.*, 1993). However, there are no reports in the literature describing how the plaque biofilm is affected by fluoride in the early stages.

Preventive programmes in orthodontic patients using fluoride mouthrinses and topical fluoride in the form of varnishes, solutions or gels have been recommended (Øgaard *et al.*, 1988b; Rolla and Saxegaard, 1990). O'Reilly and Featherstone (1987) found that toothpastes were unable to stop the development of carious lesions. They suggested that the use of a fluoride mouthrinse (0.05 per cent NaF) used daily, in addition to fluoride toothpaste (1100 ppm fluoride) could completely inhibit or even reverse demineralisation. In addition, chemical plaque control with chlorhexidine digluconate has been reported (Lundstrom *et al.*, 1980; Brightman *et al.*, 1991; Morrow *et al.*, 1992). Anderson *et al.* (1997) concluded that the use of 0.12 per cent chlorhexidine mouthrinses could be beneficial to orthodontic patients in achieving improved oral hygiene. Evidence of the efficacy of chlorhexidine in biofilms was reported by Pratten *et al.* (1998b). Their results demonstrated that chlorhexidine kills biofilms *in vitro* after pulsing twice daily over a period of 4 days.

To date, no published studies have quantitatively examined the initial attachment of bacteria to orthodontic bonding materials and the effects of chemical plaque control measures on early biofilm formation on these surfaces. This *in vitro* study aimed to quantitatively investigate the characteristics exhibited by one of the initial colonisers of the oral biofilm architecture, *Streptococcus sanguis*, on four commonly used orthodontic bonding materials in a parallel plate flow chamber. The effectiveness of two commercially available mouthrinses on *S. sanguis* biofilm formation was also evaluated.

Bacterial Strain and Media: *S. sanguis* SK36 was used in all the experiments. The organism was grown on Columbia blood agar (Oxoid, Basingstoke, Hants, UK) in order to obtain pure cultures. The cultures were prepared by inoculating a fresh colony into 20 ml tryptone soy broth (Oxoid) with yeast extract (Oxoid) and incubating in a 5 per cent carbon dioxide atmosphere overnight. The broths were centrifuged (10 minutes, 3000 rpm, 21°C) and their final density adjusted to 1×10^8 cells per ml with 0.85 per cent saline to provide a standard inoculum for flow chamber experiments. A mucin-containing artificial saliva with a pH of 6.9 was used as the growth medium for the biofilm (Pratten *et al.*, 1998a), which was based on the artificial saliva formulations by Russell and Coulter (1975) and Shellis (1978). The composition listed is per 500 ml of distilled water: 'lab-lemco' beef extract, 0.5 g (Oxoid), yeast extract, 1 g (Oxoid), proteose peptone, 2.5 g (Oxoid), mucin (hog gastric), 1.25 g (Sigma, Poole, Dorset, UK), sodium chloride, 0.175 g (Sigma), calcium chloride, 0.1 g (Sigma), potassium chloride, 0.1 g (Sigma) and 40 per cent urea, 0.313 ml (Sigma).

Bonding Materials: Discs of each of the bonding materials were prepared using a polytetrafluoroethylene (PTFE) mould. The mould consisted of five cylindrical holes, which contained PTFE plugs 5 mm in diameter. Discs of the bonding materials were prepared by firstly recessing the PTFE plug to a depth of 1 mm before filling the mould (5 x 1 mm) with material.

All experimental discs were made following the manufacturers' instructions. Fuji Ortho and Transbond were light activated with a Curing Light XL3000 (3M Unitek,

Monrovia, California, USA) for 20 seconds while Concise and Ketac Cem were allowed to bench cure for 5-7 minutes. Bovine enamel discs (Biomaterials Department, Eastman Dental Institute, London, UK) were used as control substrata. The discs were disinfected before each experiment by immersion in 1 ml of 70 per cent ethanol.

Parallel Plate Flow Chamber: Biofilms of pure cultures of *S. sanguis* were cultivated in a flow chamber (FC 71; BioSurface Technologies Corporation, Bozeman, Montana, USA). This flow chamber is specifically designed for insertion of removable discs to evaluate biofilm formation and anti-microbial testing on different surfaces. The flow chamber has a channel 0.2 mm deep x 11.4 mm wide x 40.6 mm long with a recess 2 mm deep x 25 mm long x 6 mm wide. The polycarbonate channel has a glass viewing port which allows reflected light microscopy of biofilm growth on the discs and in the flow channel. The viewing window consists of a 24 x 60 mm cover slip and is held in place by an aluminium cover plate. The cover plate also compresses the silicone rubber gasket material to provide a leak-proof cell. The flow chamber was autoclaved prior to the experiments.

Five experimental discs were placed in the flow chamber simultaneously. A peristaltic pump (Watson-Marlow, Falmouth, Cornwall, UK) was used to draw media through the chamber at a rate of 1.14 ml/minute which has been shown to represent the average salivary flow rate in orthodontic patients, corresponding to a shear rate of 11/s (Ulukapi *et al.*, 1997). There was a pulse damper between the peristaltic pump and the flow chamber to help dampen the pulsing effect of the pump. In addition to this a bubble trap for gas bubbles that might accumulate in the tubing before they enter the flow chamber was added to the apparatus.

The flow chamber was placed on the stage of a phase contrast microscope (Olympus BH-2, Tokyo, Japan) equipped with an ultra-long working distance objective (x40; Olympus ULWD-CD Plan 40 PI; producing a size of field 140 x 100 µm). A CCD camera (JVC TK-C1360B, London, UK) was attached to the microscope and images were collected onto a computer using the FlashBus MV-lite frame grabber (Integral Technologies Inc., Indianapolis, Indiana USA).

Production of Biofilms: At the start of an experiment, artificial saliva was pumped through the flow chamber for 5 minutes allowing the formation of a conditioning film on the disc surfaces. During this time, the valve of the bubble trap was opened to provide a small reservoir of liquid in the syringe barrel (2–3 ml) to act as a reservoir to take the place of the captured gas bubbles. The broth containing *S. sanguis* was then added to the artificial saliva and the experiment was allowed to proceed for 6 hours. Five hundred millilitres of artificial saliva were inoculated with 1 ml of the standardised inoculum. The culture was maintained at 37°C and stirred for the duration of each experiment. As this was a preliminary test with only a small sample size to demonstrate the functionality and applicability of the flow chamber system, this experiment was repeated once to validate the first experiment.

Corsodyl (0.2 per cent chlorhexidine gluconate; GlaxoSmithKline Consumer Health care, Brentford, Middlesex, UK) and Ortho SWIRL (0.05 per cent sodium fluoride; Ortho-Care, Bradford, West Yorkshire, UK) were used to test their effect on the early biofilm formation of *S. sanguis* on the five substratum surfaces. Before the start of the experiment, the discs were immersed in 5 ml of each agent for 1 minute before placement in the flow chamber. The production of biofilm on these chemically pre-treated discs was allowed to take place within a time frame of 6 hours.

Image Analysis: Vital staining of the *S. sanguis* biofilm was carried out using SYTO 13 live-cell nucleic acid stain (Molecular Probes Europe BV, Leiden, The Netherlands) as the biofilm was grown on non-transparent substratum surfaces. Five microlitres of the stain prepared in dimethylsulphoxide (DMSO) was pipetted onto the biofilm/disc surfaces after the 6 hour experiment. After 3 minutes incubation in the dark, at room temperature, the stained specimens were viewed under the microscope. Ultraviolet light at a wavelength of 470 nm (from a mercury-arc lamp) was used. to assist the observation of adherent bacteria on the surface of the discs. Five images per disc were captured at random. The Image (National Institutes of Health, Bethesda, Maryland, USA) analysis software programme was used to perform the image analysis. Adhering bacteria were discriminated from the background by a single grey-value threshold yielding a binary black and white image defined by the intensity levels of the original image. The percentage of the surface in the field of view covered by the biofilm was calculated by counting the pixels on threshold images to determine the proportion of pixels representing *S. sanguis* cells. An image covered a surface area of 0.012 mm^2.

Statistical Analysis: Data were analysed with the Statistical Package for the Social Sciences (Version 11.0, SPSS, Chicago, Illinois, USA). Descriptive statistics, including the mean and standard deviation of the percentage surface coverage (PSC) by *S. sanguis* biofilms, were calculated for each of the substrata. A Student's *t*-test was used to determine whether significant differences in PSC by *S. sanguis* biofilms existed between different materials. The effects of fluoride and chlorhexidine mouthrinses on biofilm formation on five substratum surfaces were analysed by univariate two-way analysis of variance (ANOVA). Dunnett's T3 test was used for *post hoc* testing. The level of significance was $P < 0.05$.

The *S. sanguis* biofilms were successfully grown in the parallel plate flow chamber consisting of single cells attached to the surfaces of bovine enamel and four orthodontic bonding materials. These cells often grow into short chains and microcolonies with recruited cells and particulate matter from the bulk fluid.

Early biofilm formation on the different bonding materials was considered low in the initial 6 hours of exposure to the medium containing *S. sanguis*. Concise showed the lowest (0.25 per cent) and Transbond the largest PSC (1.20 per cent). The Student's *t*-test revealed that Concise had a significantly lower

PSC of *S. sanguis* than the other four substrata ($P < 0.05$). Ketac Cem showed a significantly lower PSC of *S. sanguis* than Fuji Ortho and Transbond.

The PSC of biofilms on 0.05 per cent sodium fluoride-treated substrata was very low. The greatest reduction was observed on Transbond (88 per cent) and the least on Concise (52 per cent). Substrata pre-treated with 0.2 per cent chlorhexidine gluconate mouthrinse revealed a reduction in PSC of *S. sanguis* biofilms except for Ketac Cem. Interestingly, chlorhexidine was found to have no effect on the PSC of *S. sanguis* biofilm formed on Ketac Cem. The effect of chlorhexidine on Fuji Ortho was the greatest, with a 64 per cent reduction in PSC, followed by Transbond (58 per cent), bovine enamel (46 per cent) and Concise (32 per cent). Two-way ANOVA revealed a statistically significant difference between the mean values ($P < 0.05$) of PSC on the five substrata before and after treatment with the two anti-microbial mouthrinses. However, the interaction between different substrata and the two mouthrinse treatments was not shown to be statistically significant ($P = 0.056$). For comparison between the effectiveness of the two anti-microbial mouthrinses, the Dunnett's T3 multiple comparisons test indicated that 0.05 per cent sodium fluoride had a greater effect in reducing PSC by *S. sanguis* biofilms than 0.2 per cent chlorhexidine gluconate.

Biofilms were successfully grown in a flow chamber system, and image analysis facilitated comparison of the biofilms formed. Although no previous research has employed a flow chamber to study the early primary colonisation of a mono-species biofilm on orthodontic bonding materials, the use of a flow chamber to grow biofilms on surfaces of other biomaterials has been demonstrated (Bruinsma *et al.*, 2001; Van der Borden *et al.*, 2004).

Only two previous studies have investigated dental plaque accumulation around orthodontic brackets with different orthodontic bonding agents. Blunden *et al.* (1994) studied

Streptococcus mutans adherence to different orthodontic bonding agents, and found that there were statistically significant differences between these agents. In contrast, Badawi *et al.* (2003) concluded that the accumulation of *in vitro* supragingival plaque was not affected by the type of orthodontic bonding material. However, they also found that biofilms grown on fluoride releasing materials did not contain *S. mutans*. Four bonding materials were chosen in this study to represent the different categories of bonding technology used in everyday orthodontic practice (composites, GICs and RMGICs). The results of the early biofilm formation represented by PSC demonstrated significant differences between materials. Concise consistently attracted less *S. sanguis* to its surface compared with Transbond, Fuji Ortho, Ketac Cem and bovine enamel.

It is possible that as Concise is a two-paste system and its chemical cure is on-going for a number of hours after initial setting, the release of residual monomers Bis-GMA and TEGDMA from the polymerisation reaction exerts an anti-bacterial effect on the organisms on this material. The potential cytotoxicity of these monomers on bacterial growth has been postulated (Bollen *et al.*, 1997). Eliades *et al.* (1995) quantitatively investigated the release of Bis-GMA and TEGDMA residual monomer from chemically-cured and visible light-cured orthodontic adhesives. The highest monomer release was obtained from chemically-cured composite resin.

This may be due to the nature of mixing the base and catalyst pastes of the chemically-cured adhesive, which inevitably increases the bulk porosity (Fan, 1985) and thus contributes to increased bulk inhibition of polymerisation and increased potential for monomer leaching. The release of residual monomer from Transbond (light-cured composite) and Concise (chemically-cured composite) in the present study may therefore explain the different level of *S. sanguis* biofilm formed on their surfaces.

Fuji Ortho (an RMGIC) was significantly colonised by a *S. sanguis* biofilm compared with Ketac Cem (a conventional glass ionomer) in the early hours of microbial attachment. Auschill *et al.* (2002) studied, *in vivo*, the effect of dental restorative materials on dental biofilms and showed that resin-ionomer hybrids revealed a very high PSC. The results of the present investigation support their findings.

It is also worth noting that the fluoride release from RMGICs has a relatively constant rate, characterised by a substantially lower initial ion release compared with the other types of GICs (Chadwick and Gordon, 1995). The latter may be associated with the higher resistance to dissolution at the initial setting stage demonstrated by the RMGICs. These two phenomena may contribute to the difference in biofilm formation on Fuji Ortho and Ketac Cem. However, Svanberg *et al.* (1990), Hallgren *et al.* (1992) and Palenik *et al.* (1992) have shown that GICs will inhibit growth of various species of streptococci. The implication from the present study is therefore that the level of fluoride release from these orthodontic bonding materials may be too low to cause a significant inhibitory effect on early colonisation of dental plaque bacteria.

The nature of the conditioning film on the substratum surface is an important factor affecting early biofilm formation. The surface irregularities present on the experimental discs might be masked and levelled by the conditioning film (Hannig, 1999; Elliott *et al.*, 2005) and hence the protective and sheltering role of the rough surface against shear forces present in the flow chamber would be eliminated. In contrast, the conditioning film has been shown to reduce the fluoride transport out of GICs (Rezk-Lega *et al.*, 1991), a possible reason for high bacterial adherence to Fuji Ortho.

In the present study, the differential accumulation of bacteria on the five substrata pre-treated with fluoride and chlorhexidine mouthrinses has been demonstrated and a number of factors identified. The differences in: (1) the ability of the anti-microbial agents to bind to the saliva-coated (and possibly uncoated) regions of the substrata during the initial 5 minutes of the experiment when the acquired pellicle was forming; (2) their ability to remain bound to the discs (i.e. their substantivity); (3) their effectiveness, once bound to the discs, at inhibiting adherence of the bacteria; (4) the relative effectiveness of the anti-microbials at killing adherent bacteria; (5) the nature of the substratum (Deng *et al.*, 2004); and (6) the presence of other ingredients in the anti-bacterial agents (Pratten *et al.*, 1998c), may well all be contributing factors to biofilm formation on the substrata pre-treated with fluoride and chlorhexidine.

Although there was no significant interaction between different substrata and the two mouthrinse treatments ($P > 0.05$) on PSC by *S. sanguis* biofilms, this does not technically mean there is no interaction between these two effects, rather that the data did not allow a conclusion that the interaction was not due to chance. The *P*-value found was close to significance ($P = 0.056$), suggesting that there may well be an interactive 'trend' between the nature of the substratum and the effects of anti-microbial treatment. Further studies on the physicochemical properties of the substratum surfaces both with and without anti-microbial treatment are needed for clarification.

1. The parallel plate flow chamber provided a reproducible and quantitative tool to study early *S. sanguis* biofilm formation *in situ* on different orthodontic bonding adhesives. Using this technique it was also possible to evaluate the ability of anti-microbial agents to prevent biofilm formation under conditions similar to those which would exist *in vivo*. The potential to screen anti-plaque and anti-microbial agents for use in orthodontics using this approach may be of considerable value.
2. There were statistically significant differences in early *S. sanguis* biofilm formation on the various orthodontic bonding materials tested. These differences may be

attributable to the physical and chemical characteristics of the materials.

3. Fluoride mouthrinse (0.05 per cent sodium fluoride) appeared to be more effective than 0.2 per cent chlorhexidine gluconate, achieving a mean reduction of PSC of *S. sanguis* biofilm of 72 per cent compared with the control.

Biofilms and Planktonic Cells of Pseudomonas Aeruginosa have Similar Resistance to Killing by Anti-microbials

Biofilms are considered to be highly resistant to anti-microbial agents. Strictly speaking, this is not the case—biofilms do not grow in the presence of anti-microbials any better than do planktonic cells. Biofilms are indeed highly resistant to killing by bactericidal anti-microbials, compared to logarithmic-phase planktonic cells, and therefore exhibit tolerance. It is assumed that biofilms are also significantly more tolerant than stationary-phase planktonic cells. Carbenicillin appeared to be completely ineffective against both stationary-phase cells and biofilms.

Killing by this B-lactam antibiotic depends on rapid growth, and this result confirms the notion of slowgrowing biofilms resembling the stationary state. Ofloxacin is a fluoroquinolone antibiotic that kills non-growing cells, and biofilms and stationary-phase cells were comparably tolerant to this antibiotic. The majority of cells in both populations were eradicated at low levels of ofloxacin, leaving a fraction of essentially invulnerable persisters. The bulk of the population in both biofilm and stationary-phase cultures was tolerant to tobramycin.

At very high tobramycin concentrations, a fraction of persister cells became apparent in stationary-phase culture. Stationary-phase cells were more tolerant to the biocide peracetic acid than were biofilms. In general, stationary-phase cells were somewhat more tolerant than biofilms in all of the cases examined. We concluded that, at least for *Pseudomonas aeruginosa,* one of the model organisms for biofilm studies, the notion that biofilms have greater resistance than do planktonic cells is unwarranted. Tolerance to antibiotics in stationary-phase or biofilm cultures is largely dependent on the presence of persister cells.

Biofilm infections are difficult to eradicate with anti-microbial treatment, and *in vitro* susceptibility tests show considerable resistance of biofilm cells to killing. Up to 60 per cent of all human infections are caused by biofilms. It is commonly accepted that biofilms are more resistant to antibiotics than are planktonic cells. This, however, is a misconception—the biofilm literature does not point to the ability of biofilms to grow at higher concentrations of anti-microbials compared to planktonic cells.

Rather, biofilms are highly resistant to killing by bactericidal antibiotics. This should properly be referred to as phenotypic tolerance or tolerance, for short. Several factors have been suggested to account for biofilm tolerance—slow growth, the presence of an exopolysaccharide matrix that can slow the diffusion of antibiotics, and the presence of unknown resistance mechanisms. Slow growth undoubtedly contributes to resistance to killing by anti-microbials, but this is not a unique feature of biofilm cells.

With the exception of that of aminoglycosides, the exopolysaccharide matrix has not been found to notably retard the diffusion of antibiotics. For example, fluoroquinolones readily penetrate the biofilm. Multidrug resistance pumps represent a generalised resistance mechanism and have been considered as candidates for a biofilm resistance mechanism. However, at least in *Pseudomonas aeruginosa,* multidrug resistance pumps do not noticeably contribute to biofilm survival. In retrospect, this is not surprising —biofilms are not usually more resistant to growth inhibition than are planktonic cells, so there does not seem to be a need to invoke special drug resistance mechanisms.

Much progress has been made in the understanding of

biofilm development, but the molecular basis of biofilm tolerance remains elusive. In a study of *P. aeruginosa* biofilms, vast majority of cells were eliminated by fairly low, clinically achievable concentrations of fluoroquinolones, but a small fraction of persisters remained essentially invulnerable to killing. Biphasic killing kinetics revealing similar persister populations can be found in papers describing treatment of biofilms of various bacterial species with a range of bactericidal antibiotics. Persisters are largely responsible for the resistance of biofilms to killing and explain, in principle, the nature of biofilm resistance. Persisters are not mutants. Reculturing of persisters produces a wild-type population with a new population of persisters.

Testing of Planktonic and Biofilm Cell Resistance to Killing: Planktonic stationary- phase cultures were prepared by inoculating 105 cells/ml and incubating them for 18 h at 37°C with aeration. Cells were then sedimented, washed twice, and resuspended in fresh MHB at the original concentration (109 cells/ml). Planktonic logarithmic-phase cultures were prepared by inoculating 105 cells/ml and incubating them for 3 h at 37°C with aeration. For dose-dependent determination of killing, 2001 of either stationary- or logarithmic-phase cells was dispensed into microtiter plates and incubated with an antibiotic for 6 h at 37°C.

Following the challenge, the number of live cells was determined by colony counting. Biofilms were grown essentially by the method of Ceri *et al.* The device used for biofilm formation in this study is a platform carrying 96 polystyrene pegs (Nunc no. 445497) that fits as a microtiter plate lid with a peg hanging into each microtiter plate well (Nunc no. 269787). For biofilm formation, the device was placed in its original sterile tray filled with MHB and cells (104/ml) and incubated for 18 h at 37°C on a tilting shaker that provides a shearing force. After biofilms were formed on the pegs, they were washed in MHB and the device with intact biofilms was placed in a microtiter plate with fresh MHB for drug susceptibility testing. Following a 6-h incubation in the presence of an anti-microbial agent, the pegs were washed twice in MHB and the device was placed in a microtiter plate with MHB and incubated for 10 min in a water bath sonicator (Branson Ultrasonic Cleaner; Branson Cleaning Equipment Company). For each anti-microbial concentration tested, cells were collected from three parallel pegs and plated for colony counting.

Results

Several different bactericidal anti-microbials were chosen to test the relative resistance to killing of planktonic and biofilm cells—ofloxacin, a fluoroquinolone; tobramycin, an aminoglycoside; carbenicillin, a B-lactam; and peracetic acid, an oxidant. Carbenicillin is a bactericidal antibiotic that, similarly to other B-lactams, kills only rapidly growing cells. If biofilm cells were primarily slow growing, they would be expected to be resistant to killing by carbenicillin. Logarithmic-phase, stationary- phase, and biofilm cultures were challenged with carbenicillin over a wide range of concentrations, from 1.67 times the MIC (50 g/ml) to 20 times the MIC (600 g/ml). After a 6-h incubation with the antibiotic, viability was determined by colony counting.

As expected, carbenicillin produced little killing in stationary-phase cells while the majority of logarithmicphase cells were killed at 1.67 times the MIC. The amount of killing of logarithmic-phase cells dropped off significantly at concentrations above 1.67 times the MIC, indicating the presence of a persister subpopulation. This 0.1 per cent of the cells in the rapidly growing logarithmic-phase culture were invulnerable to killing by carbenicillin at 600 g/ml. Biofilm cells were resistant to killing by carbenicillin. This indicates that the biofilms used are made of slow-growing, essentially stationary-phase cells.

Unlike carbenicillin, ofloxacin can kill non-growing cells, providing a useful tool with which to examine the relative tolerance of stationary-phase and biofilm cultures. Logarithmic-phase, stationary-phase, and biofilm cultures were challenged with ofloxacin over a wide range of concentrations, from the MIC (0.5 g/ml) to 30 times the MIC (15 g/ml).After a 6-h incubation with the antibiotic, viability was determined by colony counting. The majority of cells in the three populations examined were killed by low concentrations of ofloxacin. The killing in all three cultures was distinctly biphasic, indicating the presence of persister cells. The levels of persisters were dramatically higher in the dense stationary- phase planktonic and biofilm cultures than in the logarithmic- phase cells. Note that the cell concentration of the biofilm is presented as the number of cells per peg and that the density of the biofilm is considerably higher than the density of the stationary-phase culture. Dilution of stationary-phase cells causes a drastic drop in persisters, to the level found in logarithmic- phase cells (N. Kaldalu *et al.*, submitted for publication).

The plateau at increasing antibiotic concentrations shows that persisters are essentially invulnerable to killing by a fluoroquinolone. Note that this experiment was performed with planktonic cells and biofilms. Killing of *P. aeruginosa* cells by carbenicillin. (A) Cells of logarithmic-phase, stationary-phase, and biofilm cultures were treated with carbenicillin for 6 h and then plated for colony counting. The data for biofilm cultures are plotted as CFU per peg of the biofilm device.

The limit of detection is indicated by the solid horizontal line. The standard deviation for each point was calculated with n^{3}. Biofilm, diamonds; stationary-phase cells, circles; logarithmic-phase cells, squares. (B) Data of panel A replotted as per cent survival.

6

Biomolecules in Environmental Applications

The accumulation and persistence of toxic materials in environment represents a major problem today. Various organic/inorganic chemicals are generated as by-products from various industries (e.g. petroleum and petrochemical, pulp and paper, chemical and mineral industries, etc.), which may be released into natural resources such as air, water and soil. These substances are difficult to degrade and therefore are of primary concern. Disposal of sludge generated from different industries is another major problem because of its toxic contents and needseds to be treated before its disposal. Wastewater treatment plants usually generate millions of tons of residual sludges worldwide every year.

Thus management of sludge is a major part of waste treatment, involving substantial cost and effort. For instance, estimates of 1999 sludge production in the UK were close to 1.2 million tds and were predicted to double to 2.18 million tds by the year 2006. In the developing countries, so far, almost all of the sludge is disposed off in open fields due to the shortage of appropriate disposal facilities, resulting in serious problems and leaching of heavy metals to groundwater, surface water and soil (Dubey and Juwarkar, 2001).

There are different chemical and physical methods of treatment of contaminated sites and sludges are reported, but have limited applications. Therefore, there is need of potential, environmentally compatible, non-destructive technieques or processes for environmental clean up. The basic disposal methods for such large quantities of soil and sludges contaminated with organic and inorganic pollutants are land application, landfilling, incineration, ocean dumping and lagooning. Some of the methods which are used for metal removal from sludges are as follows:

- Hydrometallurgy for metal removal
- Electrokinetic removal/electroreclamation
- Soil washing/flushing by chemicals
- Bioleaching

Use of biomolecules in environmental clean up is the recent concept. Biomolecules are the compounds or biochemicals synthesised by the living organisms which includes proteins, amino acids, organic acids, peptides, carbohydrates, enzymes, lipids, etc. These biomolecules have the capacity to interact with the polluted environment through processes like mobility, bioavailability and detoxification and the balance between mobilisation, immobilisation and detoxification varies depending on the biomolecule involved, their environment and physicochemical conditions.

Biomolecules play an important role in the environmental fate of toxic materials with a multiplicity of physicochemical and biological mechanisms effecting transformations between soluble and insoluble phases. Although, the biotechnological potential of most of these processes has only been explored at laboratory scale except bioleaching for metal extraction which has been exploited at commercial scale. Now a days, biomolecules are being exploited for broader applications and can be an efficient tool in remediation of contaminated sites and their restoration.

The different biomolecules which have potential application for treatment of contaminated soil, wastewater, water, sludges, etc., are broadly grouped as under:

1. *Cell wall substances:* Chitin and peptidoglycon,
2. *Cell membrane exudates:* Siderophores and biosurfactants,
3. *Root exudates:* Phytochelatins, organic acids, amino acids, thione related compounds, etc.,
4. *Proteins:* Metalloproteins, Metallothionins, Metal transporters, peptide dimmers, polyeptides, etc.,
5. *Enzymes:* Catalase, peroxidase, superoxide dismutase, glutathione s-trasferase, glutathione peroxidase, trasferrin, amylases, proteases, lipases, etc.

Biomolecules in Remediation

Bioremediation is the use of biological systems for the reduction of pollution from air, aquatic or terrestrial systems. Microorganisms and plants are the biological systems, which are generally exploited for bioremediation because of their capacity to degrade or modify the pollutants. Biodegradation potential of the microorganisms is the most frequently preferred bioremediation option. Microorganisms can breakdown most of the compounds for their growth and/ or energy needseds. These biodegradation processes may or may not need air. In some cases, metabolic pathways which organisms normally use for growth and energy supply may also be used to break down pollutant molecules. This mechanism is known as co-metabolism and during the process the microorganism does not benefit directly.

The use of inexpensive biological materials, such as bacteria, fungi and algae, for removing and recovering pollutants from contaminated industrial effluents has emerged as a potential alternative method to conventional techniques (Kapoor and Viraraghavan, 1995).

Every organism can be viewed as a factory of solutions that its species and progenitor species have evolved in response to problems it encountered during its struggle for survival. Therefore, the efforts are needed to exploit those mechanisms which can hold new promises to solve pollution problems. Attempts to identify organisms that can metabolise environmental pollutants have focused on organisms that have a rapid generation time (McCarty, 1993).

There has been much progress in the discovery of microorganisms that are able to degrade organic pollutants, including chlorinated compounds, polychloryl biphenyls, and polyarornatic hydrocarbons. The *in situ* application of such bacteria would involve the stimulation of pollutant-degrading biomolecule that are inherent to the area referred to as biostimulation or the introduction of toxin-degrading substance as bioaugmentation (Gadd, 1993).

Biosorption can be defined as the uptake of organic and inorganic metal species, both soluble and insoluble, by physicochemical mechanisms, such as adsorption. In living cells, metabolic activity may also influence this process because of changes in pH, EC, organic and inorganic nutrients, and metabolites. Biosorption can also provide nucleation sites for the formation of stable minerals (Beveridge *et al.*, 1989; McLean *et al.*, 2002), sorption to cellular surfaces, or accumulation within the cells via membrane transport systems of varying affinity and specificity. Once inside cells, metal species may be bound, precipitated, localised within intracellular structures or organelles or translocated to specific structures depending on the element concerned and the organism (Gadd, 1996).

Chitin

It is an important structural component of fungal cell walls and this is an effective biosorbent for radionuclides, and so are chitosan and other chitin derivatives (Tobin *et al.*, 1994). In *Rhizopus arrhizus,* Uranium biosorption involves coordination to the amine N of chitin, adsorption in the cell wall chitin structure and further precipitation of hydroxylated derivatives

(Tsezos and Volesky, 1982). Fungal phenolic polymers and melanins possess many potential metal binding sites with oxygen-containing groups including carboxyl, phenolic and alcoholic hydroxyl, carbonyl and methoxyl groups (Gadd, 1993).

Peptidoglycan

Peptidoglycan is the main cell wall compound of gram +ve bacteria. The carboxyl groups of peptidoglycan are the main binding sites in biosorption of contaminants/ metal cations. Both freely suspended and immobilised biomass from bacterial, cyanobacterial, algal and fungal species have received attention with immobilised systems appearing to possess several advantages including higher mechanical strength, easier biomass/liquid separation and, for living cells, increased metal tolerance (Brierley, 1990).

Thus, chitin and peptidoglycon binds pollutants either through adsorption and intracellular accumulation and facilitate remediation process.

Siderophores

1.3.1 Siderophores (from the Greek: "iron carriers") are defined as relatively low molecular weight, ferric ion specific chelating agents elaborated by organisms growing under low iron stress (Winkelmann, 1987). The role of these compounds is to scavenge iron from the environment and to make the mineral, which is almost always essential, available to the cell.

Siderophore is a greek phrase for iron carriers and is defined as relatively low molecular weight (500-1000 Da), produced by microorganisms or plants and accordingly called bacterial sideophores or phytosiderophores. Siderophores are produced by microorganisms or plants and accordingly called microbial siderophores or phytosiderophores.

Microbial Siderophores

Microbial siderophores are produced by naturally occurring bacteria and fungi or algae finds its application to detoxify substances hazardous to human health and/or the environment.

Siderophores though highly specific for Fe^{3+} (Leong, 1998) have been found to bind effectively to actinides and divalent metal cations such as magnesium, manganese, chromium, gallium, and radionuclides, such as plutonium. Complexation involves metabolites and siderophore that cause metal mobilisation. They play an important role in the complexation of toxic metals and radionuclides and increase their solubility (Juwarkar *et al., 1986)*. Thus, specific chelation of metals can be of great relevance to metal decontamination from various sources.

Microbial siderophores have also shown selective leaching of the crushed tailings (processed and unprocessed) naturally weathered (unprocessed) uranium ore (0.0029 per cent U by weight), kolm (0.52 per cent U by weight) and acid-leached ore (0.0099 per cent U by weight) from alum shale in the Billingen area in southern Sweden and arsenic from arsenic contaminated soil (Nair and Juwarkar, 2006).

These iron-chelating ligands that are synthesised by most microorganisms under iron-limited conditions (Neilands, 1981; Hider, 1984). Siderophores are novel structures, many containing modified amino acids not found elsewhere in nature usually are three main kinds of iron binding functional groups known as catecholate or phenolate, hydroxamate siderophores (Drechsel *et al.,* 1995; Weizman *et al.*, 1996) and carboxyalate or complexones (carboxyl and hydroxyl groups functional groups) siderophores (Dave *et al.*, 2000). The role of these compounds is to scavenge iron from the environment and to make the mineral, which is almost always essential in many biological processes such as respiration and DNA synthesis make it an essential element for almost all life forms (Wrigglesworth *et al.*, 1980). However, even though iron is the fourth most abundant element in the earth's crust, it is essentially unavailable to aerobic organisms because of its tendency to form insoluble ferric hydroxides and oxyhydroxides at neutral pH (Spiro *et al.*, 1969).

Microorganisms allow iron into the cell in dissolved form. This high affinity system has two parts; the production of an iron binding molecule, which is excreted from the cell and the elaboration of a membrane receptor molecule, which binds the iron complex and transport the metal to the cell's interior. Membrane receptor and enzymes that synthesise the iron binding molecules are coded by five genes in one operon, which is turned off when sufficient iron has been taken into the cell (Lewin, 1984).

Application in Bioremediation

Bioremediation is the use of living organisms or bioproducts, to degrade the environmental contaminants into less toxic forms. It is an option that offers the possibility to destroy or render harmless various contaminants using natural biological activity. As such, it uses relatively low-cost, low-technology techniques, which generally have a high public acceptance and can often be carried out on site.

Siderophores produced by naturally occurring bacteria and fungi or algae finds its application to degrade or detoxify substances hazardous to human health and/or the environment. The microorganisms may be indigenous to a contaminated area or they may be isolated from elsewhere and brought to the contaminated site. Contaminant compounds are transformed by biomolecules through reactions that take place as a part of their metabolic processes.

The method also has its own limitations as the range of contaminants on which it is effective is limited, the time scales involved are relatively long, and the residual contaminant levels achievable may not always be appropriate.

Phytosiderophores

Phytosiderophores (PS) are low molecular weight, organic ligands that are secreted from the plant roots of plant species (mostly from the family Graminae) as part of an Fe acquisition strategy in soils where the solubility of Fe is limited by pH. These compounds also play a role in tolerance, sequestration, and transport of inorganic and organic chemicals (Nocito *et al.*, 2002). Phytosiderophores are chelators that facilitate uptake of

Fe and perhaps other metals in grasses; they are biosynthesised from nicotianamine, which is composed of three methionines coupled via non-peptide bonds. Nicotianamine also chelates metals and may facilitate their transport. Although PS are characterised by their high affinity for Fe, the stability of some actinide-PS complexes can approach or exceed that of the Fe. Therefore, in contaminated soils PS may lead to the mobilisation of actinides by increasing the solubility and dissolution kinetics of actinides or by decreasing the absorption of actinides to mineral-oxide surfaces.

Microbe Assisted Phytoremediation (MAP) is the use of plants and their associated microbes for environmental cleanup. This technology makes use of the naturally occurring processes in which plants, phytosiderophores and microbial biomolecules secreted by rhizospheric microflora, viz., bacterial siderophores, fungal siderophores, biosurfactants, microbial membrane proteins, amino acids and organic acids degrade and sequester organic and inorganic compounds.

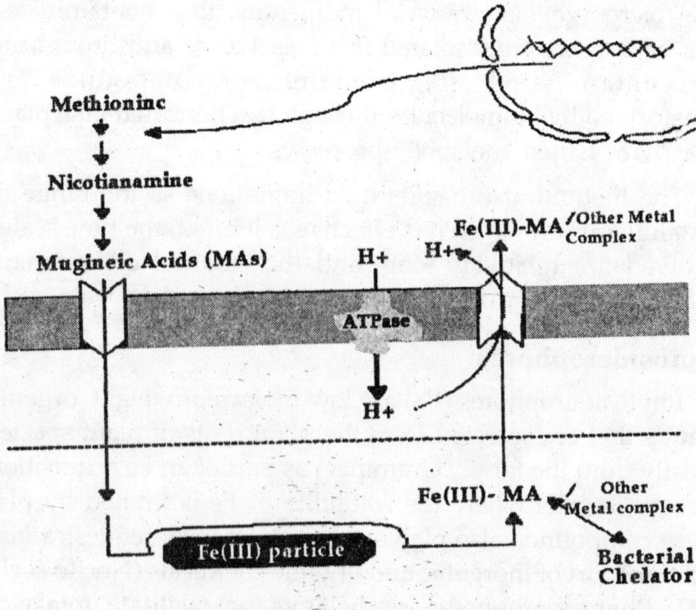

Fig: Microbe Assisted Phytoremediation pathway

Biosurfactants

Microbial surface active agents (biosurfacants) are important biotechnological products, with a wide range of applications in many industries. Their properties of interest are: (i) changing surface active phenomena, such as lowering of surface and interracial tensions, (ii) wetting and penetrating actions, (iii) spreading, (iv) hydrophylicity and hydrophobicity actions, (v) microbial growth enhancement, (vi) metal sequestration and (vii) anti-microbial action (Mei *et al.*, 1993, Dubey and Juwarkar, 2004)).

Most of the biosurfactants are high molecular-weight lipid complexes, which are normally produced under aerobic conditions. In general, biosurfactants can be classified as:

(i) Glycolipids.

(ii) hydroxylated and cross-linked fatty acids (mycolic acids).

(iii) polysaccharide-lipid complexes.

Table: Uses of Biosurfactants in Various Environmental Applications

Industry	*Application of biosurfactant*
Metals	Wetting, emulsification, lubrication and corrosion inhibition in rolling oils, cutting oils, lubricants, etc.
Paper and Pulp	Deresinification, washing, Defoaming, colour levelling, dispersing, Wetting and levelling, coating and colouring.
Waxes and polishes	Emulsify waxes, stabilise emulsions, antistat.
Petroleum production/products	Emulsify oil, disperse solids, modify rheological properties of drilling fluids for oil and gas wells.
Textiles	Detergent and emulsifier in raw wool scoring; dispersant in viscose rayon spin bath; lubricant and antistat in

Contd...

Industry	*Application of biosurfactant*
	spinning of hydrophobic filaments Wetting, penetration, solubilisation, emulsification, dye levelling, detergency and dispersion.
Agriculture	Prevent caking during storage, Wetting, dispersing, suspending of powdered pesticides and emulsification of pesticide solutions; promote wetting, spreading and penetration of toxicant.
Food and beverages	For cleaning sanitising, Improve removal of pesticides, and in wax coating, Prevent spattering due to super heat and water.
Leather	Detergent and emulsifier in degreasing.

(iv) lipoprotein-lipopeptides.

(v) phospholipids.

(vi) complete cell surface itself.

Biosurfactants, being complex organic molecules with specific functional groups, are often specific in their action (this would be of particular interest in detoxification of specific pollutants), de-emulsification of industrial emulsions, specific cosmetic, pharmaceutical, and food applications.

Mineralisation and removal of organic and inorganic pollutants using biomolecules is an emerging area. Few bench scale studies have been conducted *to* evaluate the effect of the biosurfactants on biodegradation of organic pollutants like hydrocarbons like *n*-paraffin, *n*-alkaned, octadecanes, etc. (Noordman *et al.*, 2002; Rahman *et al.*, 2003), polycyclic aromatic hydrocarbon like phenanthrene, naphthalene in soils and in liquids (Deschenes *et al.*, 1994; Vipulanandan *et al.*, 2000; Rahman *et al.*, 2002) and chlorinated hydrocarbons like the pesticides, DDT, 2,4-D and 2,4,5-T, plasticisers, pentachlorophenol, polychlorinated biphenyls and others (Robinson, *et al.*, 1996; Fiebig *et al.*, 1997; Meta-Sandoval *et al.*, 2001).

Use of biosurfactant facilitates the removal of metals from contaminated soil. Lead-contaminated soil was investigated for cleanup using biosurfactant (Neilson *et al.*, 2003). An investigation involving the addition of surfactant to remediate cadmium-contaminated soils was performed to determine the optimal surfactant enhanced remediation system (Doong *et al.*, 1998). Column experiments were conducted to evaluate the feasibility of using a rhamnolipid foam to remove heavy metals (Cd and Ni) from a sandy soil contaminated (Wang *et al.*, 2004)

The feasibility of using a biodegradable surfactant, surfactin from Bacillus subtilis, for the removal of heavy metals from a contaminated soil and sediments was evaluated (Mulligan *et al.*, 1999). Remediation of metal-contaminated soil and sludge using biosurfactant was studied by (Maier *et al.*, 2001). Mining residues, in general and metal ores, in particular, contains heavy metals within the rock. These heavy metals, including copper (Cu) for which there is an increasing demand around the world, and are very harmful to humans and, as such, are a serious problem for the environment. Dahrazma *et al.*, in 2004 reported the use of rhamnolipid biosurfactant to extract copper from an oxide residue with 8,950 mg copper per kg ore up to 42 per cent in 6 days.

There is an increased interest in removal of metals from water streams from industrial discharge and mining operations. Removal of lead from waste waters using biosurfactant has been studied by Kim *et al.*, 2006. It was found to be more effective that Sodium Dodecyl Sulphate (SDS) and Triton X-100.

Biosurfactant in the biological control of zoosporic, bacterial and fungal plant pathogens has been indicated that they can be exploited commercially (Stanghellini et. al, 1997). Zoosporic fungi are among the most destructive root pathogens in recirculating hydroponic cultural systems, and zoospores have been implicated as the primary, if not sole, infectious propagule responsible for the spread of these pathogens via the recirculating nutrient solution. In vivo experiments employing cucumbers as the susceptible host and Pythium aphanidermatum as the root pathogen demonstrated the efficacy of surfactants in the control of root disease caused by this fungus. In the absence of a surfactant, all plants within a recirculating hydroponic unit were killed within 5 to 6 weeks following hypocotyl inoculation of a single plant in the hydroponic unit (Stanghellini *et al.*,1996). A lipopeptide from B. *thuringiensis* CMB26 and found to be similar to the lipopeptide fengycin. This lipopeptide can function as a biocontrol agent, and exhibits fungicidal, bactericidal, and insecticidal activity. Compared with surfactin and iturin, the lipopeptide from *B. thuringiensis* CMB26 showed stronger anti-fungal activity against phytopathogenic fungi. This lipopeptide is a candidate for the biocontrol of pathogens in agriculture (Kim *et al.*, 2004).

Phytoremediation, the use of plants and their associated microbes for environmental cleanup, has gained acceptance in the past 10 years as a cost-effective, non-invasive alternative or complementary technology for engineering-based remediation methods. Plants can be used for pollutant stabilisation, extraction, degradation, or volatilisation. Transgenic approaches are being used to enhance these processes (Pilon-Smits E., 2005). A strategy originally intended for phytoremediation of hydrocarbons may produce crops capable of degrading excess agricultural chemicals. Rhamnolipids, normally produced by the bacterium *Pseudomonas aeruginosa,* may prove useful in designing crops with built-in environmental stewardship genes. Initial field studies convincingly support using rhamnolipid applications to facilitate the degradation of complex hydrocarbons in soil. A recent investigation by Nickolai Kartel, of the Nations Academy of Sciences in Belarus, has successfully reduced soil hydrocarbons using transgenic plants that secrete rhamnolipids (Brychkova *et al.* (Submitted 2003).

Mineralisation and removal of organic and inorganic

pollutants using biomolecules is an emerging area. Few bench scale studies have been conducted to evaluate the effect of the biosurfactants on biodegradation of organic pollutants such as hydrocarbons like *n*-paraffin, *n*-alkaned, octadecanes, etc., polycyclic aromatic hydrocarbon like phenanthrene, naphthalene in soils and in liquids (Bodour *et al.*, 2002) and chlorinated hydrocarbons like the pesticides, D.D.T, 2,4 - D and 2,4,5 - T, plasticisers, pentachlorophenol, polychlorinated biphenyls and others.

Use of biosurfactans facilitates the removal of metals from contaminated soil. Lead-contaminated soil was investigated for cleanup using biosurfactant. An investigation involving the addition of surfactant to remediate cadmium-contaminated soils was performed to determine the optimal surfactant enhanced remediation system. Column experiments were conducted to evaluate the feasibility of using a rhamnolipid foam to remove heavy metals (Cd and Ni) from a contaminated sandy soil. Biosurfactant in the biological control of zoosporic, bacterial and fungal plant pathogens has been indicated that they can be exploited commercially for pest control also (Stanghellini et. al, 1997, Nair and Juwarkar, 2006).

Peptides, Proteins and other Exudates

A range of specific and non-specific metal-binding compounds are produced by living organisms. Non-specific metal-binding compounds range from simple organic acids and alcohols to macromolecules, such as polysaccharides, humic and fulvic acids (Gadd, 1993). Extracellular polymeric substances (EPS), a mixture of polysaccharides, mucopolysaccarides and proteins, are produced by bacteria, algae and fungi and also bind potentially with toxic metals (Gadd, 1996).

Extracellular polysaccharides can also adsorb or entrap particulate matter such as precipitated metal sulphides and oxides (Grill *et al.*, 2001). Floating cyanobacterial mats are used to remove metals from waters. The metal-binding process being due to large polysaccharides (>200,000 Da; Gadd, 1996).

Metalloproteins

Specific, low molecular weight (6000-10,000 Da) metal-binding proteins, termed as metalloproteins, are produced by animals, plants and microorganisms in response to the presence of toxic metals (Rauser, 1995). Metal-binding proteins, phytochelatins and related peptides, contain glutamic acid and cysteine at the amino-terminal position, and have been identified in plants, algae and several microorganisms (Rauser, 1995).

Eukaryotic metallothioneins and other metal-binding peptides have been expressed in *E. coli. In vivo* expression of metallothioneins provides a means of designing biomass with specific metal-binding properties (Winkelmann *et al.*, 1987). The availability of different metal ions with various redox states, ligands, and geometries about the metal ion allows for fine-tuning of the activity of proteins at the highest level.

Phytochelatins

Many plants and some fungi, respond to excess metals by producing a class of small metal-binding peptides called as phytochelatins (PCs) (Rauser, 1995) and are derived from glutathione, PCs have the general structure (y-glutamyl-cysteine), as indicated by the presence of the y-glutamic acid linkages in the peptide chain. The first step in the synthesis of PCs from glutathione is the joining of a cysteine residue to glutamic acid and this is catalysed by y-glutarnylcysteine synthetase (Jackson *et al.*, 1987). Following this glutathione synthetase catalyses the reaction of the addition of glycine to y-glutamylcysteine.

Lastly, PC synthase, an enzyme constitutively present in plant cells, elongates the peptide to produce a peptide consisting of 3-y-glutarnylcyteine residues and a glycine (Grill *eit al*, 1989). Presumably, metal ions activate PC synthase to initiate PC production. The resultant PCs in turn eventually deactivate PC synthase by chelating the heavy metal cofactors (Grill *et al.*, 1989).

Root Exudates

Contaminant transport experiments in the presence of plants have shown both increased and decreased leaching of metals compared to plant-free systems. Organic acids such as citrate or oxalate exudated from plant roots into the rhizosphere may affect the pollutant behaviour in soils and water.

An exclusion mechanism for metal tolerance in plants renders the metals present in the ambient environment less bioavailable. It has been postulated that if certain metal-chelating compounds could be excreted by plant roots then the metals could, and would, no longer enter into the root apex (Kochian, 1995). Indeed, recent experimental evidences have indicated that metal-binding ligands which are released into the rhizosphere by certain plant species, make them resistant to metals and can survive in contaminated sites (Miyasaka *et al.*, 1991).

Biomolecules in Pollutant Degradation

Bacteria degrade compounds by producing very specific enzymes. Bacteria are extremely efficient "enzyme factories". They recognise the organics present in an environment and respond by producing the enzymes required to degrade those specific organics. The breakdown of waste is a complicated metabolic pathway in which each step requires a specific enzyme. For example, a triglyceride is initially cleaved by lipase into a glycerol and three fatty acids. The glycerol and each fatty acid are then broken into smaller components by specific enzymes until eventually these compounds are reduced to carbon dioxide and water.

Microbial enzymes are of recent interest and enabling new technologies for processing pulps and fibres. Xylanases reduce the amount of chemicals required for bleaching; cellulases smooth fibres, enhance drainage, and promote ink removal; lipases reduce pitch; lignin-degrading enzymes remove lignin from pulps. There are other classes of enzymes such as cellulose, xylan and pectin, etc., which aids in degradation of complex compounds.

Biodegradation of Cellulose

Four classes of enzymes are involved in the biodegradation of cellulose. Endoglucanases (EC 3.2.1.4) hydrolyse cellulose to glucooligosaccharides. Cellobiohydrolases .(EC 3.2.1.91) release cellobiose from crystalline cellulose. Glucosidases (EC 3.2.1.21) degrade the oligosaccharides to glucose. Exoglucanases release glucose from cellulose and glucooligosaccharides.

Degradation of Xylan

The biodegradation of the xylan backbone depends on two classes of enzymes. Endoxylanases (EC 3.2.1.8) are able to cleave the xylan backbone into smaller oligosaccharides, which can then be degraded further to xylose by xylosidases (EC 3.2.1.37). Both classes of enzymes, as well as their encoding genes, have been characterised from many organisms.

Degradation of Pectin

The structural differences between the main chain of the hairy and smooth regions of pectin have implications for the enzymes involved in the degradation of these regions. The backbone of the smooth region can be hydrolysed by pectin lyases (EC 4.2.2.10), pectate lyases (EC 4.2.2.2), and polygalacturonases (EC 3.2.1.15 and EC 3.2.1.67). In *Aspergillus*, families of genes encoding these types of enzymes have been identified. Several classes of enzymes are involved in the degradation of the hairy-region backbone. The pectin main-chain-degrading enzymes can be divided into hydrolases and lyases. Six types of hydrolases have been identified in aspergilli. Several endopolygalacturonases are produced that all cleave within the pectin smooth region, whereas exopolygalacturonases cleave at the non-reducing terminal end of this region.

Enzymes Involved in the Degradation of Plant Cell Wall Polysaccharides.

In contrast to the enzymes which act on the main chain of plant cell wall polysaccharides, accessory enzymes act on the substituents or the side chains of these structures. Some of these enzymes act on linkages between a main-chain residue and a substituent, whereas other enzymes cleave internal or terminal linkages of side chains.

D-Xylosidases

D-Xylosidases can release linked xylose residues from xyloglucan. Only a limited number of xylosidases has been characterised from many microorganisms. These enzymes are all highly specific for linked xylose residues but differ with respect to the type of glycoside they can hydrolyse.

U- and D-Galactosidases.

The removal of D-galactose residues from plant cell wall polysaccharides requires the action of galactosidases (EC 3.2.1.22) and galactosidases (EC 3.2.1.23). Galactosidases release terminal galactose residues from the galactan side chains of pectins. Galactosidases are involved in the degradation of galacto (gluco) mannan, removing galactose from the mannose residues of the backbone. The presence of terminal linked galactose residues in certain galactoglucomannans suggest that both galactosidases may play a role in the degradation of these polysaccharides.

Endo and Exogalactanases

The galactan side chains of pectin are hydrolysed by endogalactanases (EC 3.2.1.89), exogalactanases, and galactosidases. Endogalactanases are able to hydrolyse the galactan polysaccharides, resulting in the liberation of galactobiose and galactose. Production of endogalactanases was observed on beet pulp, soybean, and locust bean gum. Differences between the enzymes exist with respect to their ability to hydrolyse linkages between galactose residues.

Biomolecules Synthesised under Stress Conditions and their role in Resistance to Pollutants.

Most organisms are exposed to toxic substances and have developed detoxification systems to transform, metabolise and eliminate such compounds from their cells. Remarkably, a common pathway exists in most organisms for the detoxification of electrophilic compounds, which is mediated by three groups of reactions (Foyer, 1987). Phase-I, transformation enzymes such as cytochrome-P450 monooxygenases and peroxidases introduce functional groups onto xenobiotic substrates. Phase-II, conjugation enzymes such as UDP-glucosyl-transferases and GSTs utilise these functional groups as a site of further conjugation, usually resulting in less toxic and more water-soluble conjugates.

Most organisms including plants, mammals, invertebrates, bacteria, etc., encode multiple isoforms of both phase I and phase II enzymes, each with their own substrate specificity. Phase-III, compartmentation enzymes, for example in plants, ATP-dependent membrane pumps, recognise and transfer conjugates across the membranes into the vacuole or to the apoplast (Halliwell and Foyer, 1978).

In plant cells, chloroplasts, mitochondria, and peroxisomes are intracellular generators of activated oxygen species such as H_2O_2, superoxide radicals, hydroxyl radicals, and singlet oxygen. Some of these activated oxygen species are highly reactive and in the absence of protective mechanisms, can produce damage to cell structure and function (Halliwell and Foyer, 1978). Under non-stressing conditions the anti-oxidative defence system provides adequate protection against activated oxygen species (Nakano and Asada, 1980).

To minimise the oxidative damage, plants have evolved defence systems involving anti-oxidants and protective enzymes that either prevent the formation of or scavenge reactive oxygen species (ROS). Super oxide dismutase (SOD) catalyses the dismutation of superoxide radicals to molecular oxygen and H_2O_2, thus playing a key role in this defence mechanism (Ho *et al.*, 1992). H_2O_2 scavenging is accomplished by catalase, various peroxidases, and the ascorbate-glutathione cycle, a series of coupled redox reactions involving four enzymes, ascorbate peroxidase (APX), monodehydroascorbate reductase (MDHAR), dehydroascorbate reductase (DHAR), and glutathione reductase (GR) (Foyer and Halliwell, 1976; Nakano and Asada, 1980). Susceptibility to oxidative stress depends on the overall balance between factors that increase oxidant generation and those cellular components that exhibit an anti-oxidant capability (Foyer *et al.*, 1997).

Peroxidases (POX)

These are abundant in plants and are characterised according to their isoelectric points as acidic or basic peroxidases. They are ubiquitous enzymes involved in the detoxification of H_2O_2 to water at the expense of electrons obtained from reducing substrates. Plant peroxidases are localised mainly in cell wall and in vacuoles and are associated with the non-specific oxidative polymerisation of phenolic units in the cell wall to make lignin. Plant roots exude high concentrations of peroxidases into soil, particularly in response to water stress and plant die-back and in response to chemical stress (Nakano and Asada, 1980).

Guaiacol

It is a substrate metabolised by a wider array of POX, thus it may serve as a marker for general POX activity.

Glutathione

Glutathione is found in the vast majority of prokaryotic and eukaryotic cells, where it often represents the major pool of non-protein reduced sulphur (Kunert and Foyer, 1993). The reduced form of glutathione (GSH) is a tripeptide thiol with the formula c-glu-cys-gly. Owing to its redox-active thiol group, GSH has often been considered to play an important role in defence of plants and other organisms against oxidative stress (Alscher, 1989; Grant *et al,* 1996). In all cells where GSH is found, the reduced tripeptide form exists interchangeably with the oxidised form (glutathione disulphide: GSSG). While glutathione reductase (GR) uses NADPH to reduce GSSG to GSH, various free radicals and oxidants are able to oxidise GSH to GSSG.

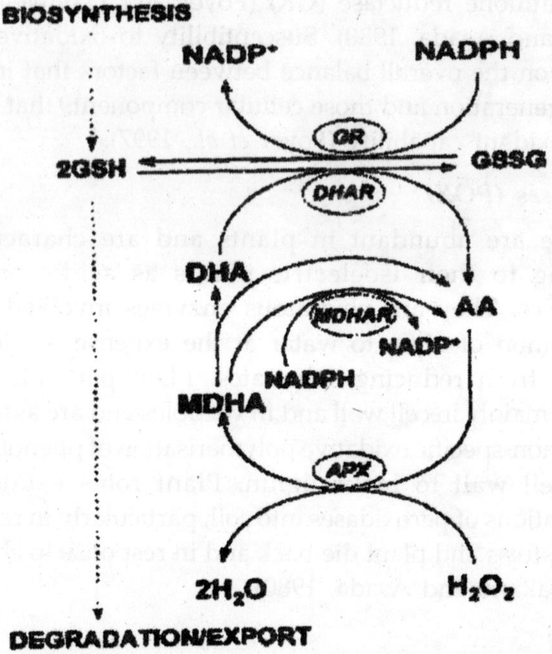

***Fig.** Schematic representation of biosynthesis and interactions of different forms of glutathione in removal of* H_2O_2

Glutathione is widely used as a marker of oxidative stress to plants, although its part in plant metabolism is a multifaceted one (Grill *et al.*, 2001). As it is a non-protein sulphur-containing tripeptide, glutathione acts as a storage and transport form of reduced sulphur. Sulphur compounds such as SO_2 and H_2S, although naturally occurring as atmospheric trace gases, are major anthropogenic air pollutants. Their toxic effects to plants have been the subject of many studies without being fully clarified, but they have well-documented effects on plant glutathione metabolism (De Kok and Tausz, 2001).

Glutathione is related to the sequestration of xenobiotics and heavy metals and is also an essential component of the cellular anti-oxidative defence system, which keeps reactive oxygen species (ROS) under control (Noctor and Foyer, 1998). Anti-oxidative defence and redox reactions play a central role in the acclimatisation of plants to their environment, which made glutathione a suitable candidate as a stress marker. Multiple roles of glutathione in plant metabolism has been shown in figure.

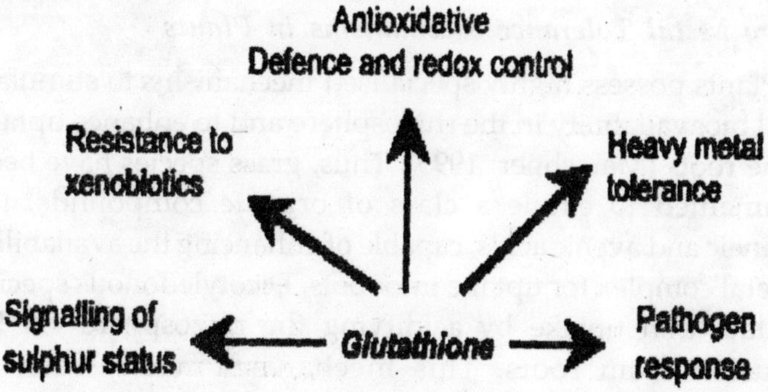

Fig. Multiple roles of glutathione in plant metabolism

Glutathione S-transferases (GSTs)

These are the enzyme class that catalyses the conjugation of the tripeptide glutathione (GSH, c glutamylcysteinylglycine) to a variety of hydrophobic, electrophilic, and usually cytotoxic substrates. GSTs are most often thought of as detoxification enzymes and indeed were first discovered for their ability to metabolise a wide range of toxic exogenous compounds via GSH conjugation (Mannervik and Danielson, 1988). The key role played by GSTs is their ability to inactivate toxic compounds (Mitra *et al.*, 1992). GSTs catalyse the nucleophilic attack of the sulphur atom of glutathione at the electrophilic centre of a wide variety of substrates.

Biomolecular Mechanism involved in Heavy Metal Resistance.

As a process for immobilising metals, precipitation of metals as sulphides has achieved large-scale application. Continued exploitation of other biological processes will undoubtedly depend on a number of scientific and economic factors. Finally, it should be emphasised that this area of research also provides understanding of the biogeochemistry of metal(loid) cycling in the environment and the central role of microorganisms in affecting metal mobility and transfer between different biotic and abiotic locations.

Heavy Metal Tolerance Mechanisms in Plants

Plants possess highly specialised mechanisms to stimulate metal bioavailability in the rhizosphere and to enhance uptake by the roots (Marschner, 1995). Thus, grass species have been documented to exude a class of organic compounds, i.e. mugineic and avenic acids, capable of enhancing the availability of metal complex for uptake into roots. Dicotyledonous species facilitate iron uptake by acidifying the rhizosphere via H^+ extrusion from roots. This mechanism mostly seen in hyperaccumulator species.

Hyperaccumulators are species capable of accumulating metals at levels 100-fold greater than those typically measured in shoots of the common non-accumulator plants. Thus, an hyperaccumulator can concentrate more than 10 ppm Hg; 100 ppm Cd; 1000 ppm Co, Cr, Cu, and Pb; 10 000 ppm Zn, and Ni. To date, approximately 400 plant species from at least 45 plant families have been reported to hyperaccumulate metals (Baker *et al.*, 2000).

Table: Different tolerant mechanisms and related biomolecules

Mechanism	*Biomolecule*
Induction of proteins	Metallothionein Heat-shock protein Phytochelatins Ferritin Transferrin
Induction and activation of anti-oxidant enzymes	Superoxide dismutase Catalase Glutathiono peroxidase
High turnover of organic acids	Phytate, malate, citrate, oxalate, succinate, aconitate, α -ketoglutarate

In general, in order to survive in metal contaminated

environment, plants must have developed efficient and specific mechanisms by which metals are taken up and transformed into a physiologically tolerable form, providing the essential elements for the metabolic functions of plants. On the other hand, excess of these essential elements, or of toxic metals that do not play a role in plant metabolism have to be metabolically inactivated. The mechanisms proposed for metal detoxification and hyperaccumulation within the plant involve chelation of the metal cation by ligands and/or sequestration of metals away from sites or metabolism in the cytoplasm, notably into the vacuole or cell wall.

Other possible adaptive responses include activation of alternative metabolic pathways less sensitive to metal ions, modification of enzyme structure or alteration of membrane permeability by structural reorganisation or compositional changes. There is, however, insufficient evidence to support that the latter mechanisms of cellular tolerance are important in hyperaccumulator plants.

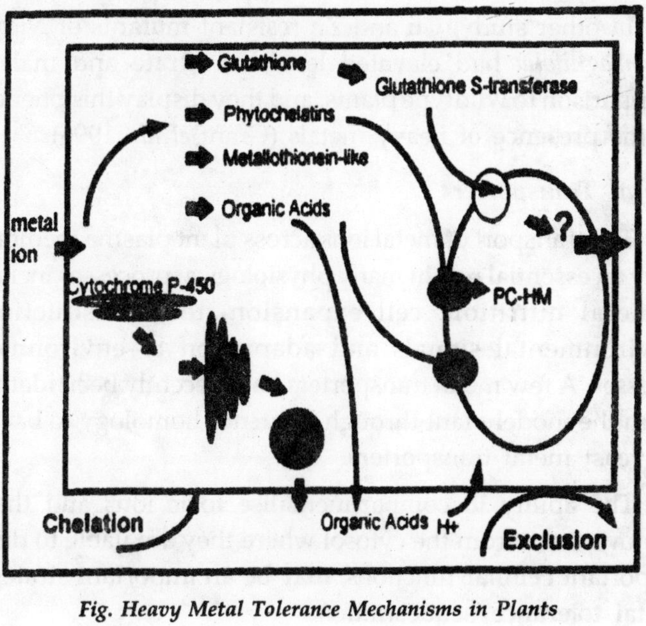

Fig. Heavy Metal Tolerance Mechanisms in Plants

Table: Functions of Metal-Biomolecule complex Organic Acids

Protein	*Metal*	*Function*
Transferrin	Cu. Fe. Mn. V	Transport
Ferritin	Fe	Storage
Cerulopiasmin	Cu	Transport
Albumin	Cu. Ni. Zn	Transport
α -Macroglobin	Zn	Transport
Metallothionein	Cd, Cu, Hg, Zn	Storage

It has long been observed that organic acids chelate heavy metals, but there is conflicting evidence regarding the role of organic acids in the metal accumulating ability of hyperaccumulators. In numerous Ni accumulating species, such as *Serberia acziminata. Homalium fiancii, Homalium kanaliese. Hybanthus ausrroculedonicus* and *Kvbanthzis caledonicirs*, a correlation between organic acids and heavy metal tolerance has been reported.

In other study, Cu and Zn resistant mutants of *Nicotiuna plumbagnifolia* had elevated levels of citrate and malate in comparison to wildtype plants, and they display this phenotype in the presence of heavy metals (Grant *et al.*. 1996).

Metal Transporters

The transport of metal ions across plant plasma membranes play an essential role in many physiological processes including mineral nutrition, cell expansion, the transduction of environmental signals and adaptation to environmental stresses. A few metal transporters have recently been identified from the model plant through sequence homology to bacterial or yeast metal transporters.

The ability to compartmentalise toxic ions and thereby remove them from the cytosol where they are liable to disrupt important cellular functions, may be an important strategy of metal tolerance/sequestration.

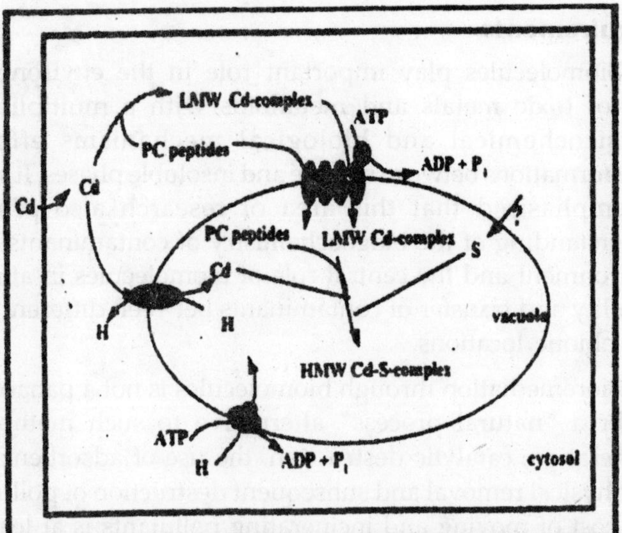

Fig The Yeast Vacuole Transporter, HMT, and its Role in Heavy Metal Tolerance

Metals can also be bound by the thiol-rich peptides GSH and PCs, or by the Cys-rich MTs. Chelated metals in roots may be stored in the vacuole or exported to the shoot via the xylem. Organics may be conjugated and stored or degraded enzymatically. Chelation in roots can affect phytoremediation efficiency as it may facilitate root sequestration, translocation, and/or tolerance. Root sequestration may be desirable for phytostabilisation whereas export to xylem is desirable for phytoextraction.

Inorganic chemicals require membrane transporter proteins to be exported from the root endodermis into the root xylem. Some inorganics are chelated during xylem transport by organic acids (histidine, malate, citrate), nicotianamine, or thiol-rich peptides. For most inorganics it is still unclear via which transporter proteins they are exported to the root xylem and to which if any chelators they are bound during transport. Better knowledge of the transporters and chelators involved in translocation of inorganics would facilitate the development of transgenics with more efficient phytoextraction capacity.

Requirements

Biomolecules play important role in the environmental fate of toxic metals and metalloids with a multiplicity of physicochemical and biological mechanisms effecting transformations between soluble and insoluble phases. It should be emphasised that this area of research also provides understanding of the biogeochemistry of contaminants in the environment and the central role of biomolecules in affecting mobility and transfer of contaminants between different biotic and abiotic locations.

Bioremediation through biomolecules is not a panacea but rather a "natural process" alternative to such methods as incineration, catalytic destruction, the use of adsorbents, and the physical removal and subsequent destruction of pollutants. The cost of moving and incinerating pollutants is at least ten times that of *in situ* biological treatment. By integrating proper utilisation of natural or modified microbial capabilities with appropriate engineering designs to provide suitable growth environments, bioremediation can be successful in the field.

However, a gap exists between advances in laboratory research and commercial field applications. Two major factors responsible for this gap are the lack of (1) a sufficient knowledge base to accurately predict pollutant degradation rates and fates and (2) sites designated as field research centres for bioremediation research and technology demonstrations. Laboratory and field studies have documented the potential use of biomolecules for bioremediation. However, the physiologic potential of microbial populations to remediate environments of relevant size, heterogeneity, and variability has not been adequately tested. Successful application of bioremediation techniques must address both the heterogeneous nature of many contaminated waste sites and the complexity of using living organisms.

There has been progress in overcoming some of the barriers that have impeded bioremediation from being successfully applied in the field. Scientists have begun to search for organisms with better biodegradation kinetics for a variety of contaminants within broad environmental habitats. Studies examining extremophiles could result in using biomolecules *in situ* that have a high tolerance for organic solvents and alkaline soils or waters, and that function at high temperatures for more efficient *ex situ* activity in bioreactors.

The research programme areas, listed below, need fundamental understanding of how to apply biomolecular technologies to engineering activities.

1. Improve site characterisation technology.
2. Create a "Biomolecular remediation Field Initiative" to expand field experiments and collect and disseminate bioremediation performance data.
3. Develop new approaches to performance evaluation in order to better determine the extent and rate of cleanup.
4. Research processes to identify biomolecular capability of biotransformation and develop methods for delivery.
5. Develop better models of bioremediation through exploitation of field results to predict and guide future applications.

To meet society's needseds and the challenges in Environmental policies, research in biomolecules must develop methodologies and tools to minimise negative environmental impacts, reclaim and wisely use finite resources.

Public concern and legislation are nowadays demanding better environmental control. In addition, the increase in the number of analytes to monitor requires more suitable analytical methods. For conventional "off-site" analysis, samples need to be sent to a laboratory for testing. Conventional methods are actually reaching the highest accuracy with low detection limits, but are expensive, time-consuming, and require the use of highly trained personnel. The current tendency to carry out field monitoring has driven the development of biosensors as new analytical tools able to provide fast, reliable, and sensitive measurements with lower cost; many of them aimed at on-site analysis. In this sense, biosensors would not compete with official analytical methods, but they can be used both by regulatory authorities and by industry to provide enough information for routine testing and screening of samples Biosensors are defined as analytical devices incorporating a biological material, a biologically derived material, or biomimic, intimately associated with or integrated within a physicochemical transducer or transducing microsystem. Biosensors should be distinguished from a bioassay or a bioanalytical system, which require additional processing steps, such as reagent addition and where the assay design is permanently fixed in the construction of the device.

Biosensors are being developed for different applications, including environmental and bioprocess control, quality control of food, agriculture, military, and, particularly, medical applications. In fact, most of the commercially available biosensor systems are applied in the clinical and pharmaceutical markets. Accordingly, most research and development has been devoted to this area. In the food industry, the detection of contaminants, verification of product content, monitoring of raw materials conversion, and product freshness are areas of potential biosensor application. The beer industry has already identified ways for improving and controlling their products through the use of biosensors.

Biosensors can be also a defence tool through the early detection of hazardous materials such as germs or chemical warfares. The detection of illicit drugs and explosives, with airport security purposes, is also a major area of research in biosensors. For environmental control and monitoring, biosensors can provide fast and specific data of contaminated sites. They offer other advantages over current analytical methods such as the possibility of portability and working on-site and the ability of measuring pollutants in complex matrices with minimal sample preparation. On the other hand, biosensors offer the possibility of determining not only specific chemicals, but also their biological effects, such as toxicity or endocrine-disrupting effects, which are sample information of great interest.

Biosensor Configurations

Biosensors are usually classified into various basic groups, according to the signal transduction and to the biorecognition principles. On the basis of the transducing element, biosensors can be categorised as electrochemical, optical, piezoelectric, and thermal sensors. The electrochemical biosensors, and among them the amperometric and the potentiometric ones, are the best described in the literature; those based on optical principles are the next most commonly used transducers. In fact, most catalytic biosensors are based on electrochemical methods, whereas affinity biosensors have generally proved more amenable to optical detection methods. The various types of optical transducers exploit properties such as simple light absorption, fluorescence/phosphorescence, bio/chemiluminescence, reflectance, Raman scattering, and refractive index. Surface plasmon resonance (SPR) is another common transduction mechanism whose main advantage over most optical biosensors is that the analyte presence can be determined directly, without the use of labelled molecules. Finally, cantilever biosensors are an emerging group of biosensors, which are based on the bending of silicon cantilevers caused by the adsorption of target molecules onto the cantilever surface, where receptor molecules are immobilised.

According to the biorecognition principle, biosensors are classified into immunochemical, enzymatic, non-enzymatic receptor, whole-cell, and DNA biosensors. Immunosensors present the advantages of sensitivity and selectivity inherent to the use of immunochemical interactions. Limitations are the troubles derived from the regeneration of the immunosurface, and cross-reactivity, although a certain degree of cross-reactivity is often desirable in order to determine different congeners of the same family.

Enzymes are suitable to act as recognition elements because

of their specificity and the wide range of enzymes available. In general, enzymatic biosensors are based on the selective inhibition of specific enzymes by different classes of compounds. The decrease of activity of the immobilised enzyme in the presence of the target analyte is frequently used for its quantification. Biosensors based on natural receptors can be built by integrating the specific receptor within a membrane and by coupling it to a transducing device.

These natural receptors are proteins of non-catalytic or non-immunogenic origin, which span cell membranes and can specifically bind certain compounds. The binding signal is detected as a structural change or an associated enzyme activity. Whole cells of living organisms, such as bacteria, yeast, fungi, plant and animal cells, or even tissue slices have been used as the recognition component by interrogating their general metabolic status. These biosensors are useful to determine whether a substance is toxic to certain cells. Another application of whole-cell biosensors is the determination of the "biological oxygen demand" (BOD).

In the case of DNA biosensors, two strategies are applied to detect pollutants: one is the hybridisation detection of nucleic acid sequences from infectious microorganisms, and the other one, the monitoring of small pollutants interacting with the immobilised DNA layer (drugs, mutagenic pollutants, etc.).

One key step in the development of biosensors is the immobilisation of the biological component at the transducer surface. The immobilisation procures both the stabilisation of the biomaterial and the proximity between the biomaterial and the transducer. The immobilisation methods most generally employed are: physical adsorption at a solid surface, cross-linking between molecules, covalent binding to a surface, and entrapment within a membrane, surfactant matrix, polymer or microcapsule. In addition to these conventional methods, sol-gel entrapment, Langmuir–Blodgett (LB) deposition, electropolymerisation, self-assembled biomembranes and bulk modification have been also used.

Toxicity

In environmental pollution monitoring, it is becoming a general opinion that chemical analysis by itself does not provide sufficient information to assess the ecological risk of polluted waters and wastewaters. In the European Union, along with more stringent demands for water treatment (Council Directive 91/271/EEC), industrial and urban wastewater effluents shall reach certain limits of non-toxicity before the effluent can be discharged into the environment. Thus, much effort has been made during the last years to develop and use different bioassays and biosensors for toxicity evaluation of water samples.

Whole organisms are used to measure the potential biological impact (toxicity) of a water or soil sample. That is the case of the toxicity assays Microtox (Azure, Bucks, UK), or ToxAlert (Merck, Darmstadt, Germany). These systems are based on the use of luminescent bacteria, *Vibrio fischeri*, to measure toxicity from environmental samples. Bacterial bioluminescence has proved to be a convenient measure of cellular metabolism and, consequently, a reliable sensor for measuring the presence of toxic chemicals in aquatic samples. Some bioassay methods are integrated now in biosensors such as the Cellsense, which is an amperometric sensor that incorporates *Escherichia coli* bacterial cells for rapid ecotoxicity analysis. It uses ferricyanine, a soluble electron mediator, to divert electrons from the respiratory system of the immobilised bacteria of a suitable carbon electrode.

The resulting current is, thus, a measure of bacterial respiratory activity, and the perturbation by pollutants can be detected as a change in the magnitude of the current. Cellsense has been applied to investigate the toxicity of 3,5-dichlorophenol and other phenols in wastewater, for the determination of non-ionic surfactants and benzene sulphonate compounds, for the analysis of wastewater treatment works (WWTW) influent and effluent, and for the toxicity testing of wastewaters and sewage sludge. Moreover, Cellsense has been proposed as one of the newer rapid toxicity assessment methods within the direct toxicity assessment (DTA) demonstration programme of the UK Environmental Agency.

Most environmental biosensors have focused on bacterial systems while eukariotic biosensors are rare; even more rare is the use of mammalian cells. The mammalian cell, which is more complex than bacteria, can give a more sensitive response when compared to bacteria while also responding to the estrogenic effects of chemicals. A recombinant fluorescent Chinese Hamster Ovary cell line, utilising a fluorescent reporter system, was used to monitor various toxicants, especially endocrine-disrupting compounds (EDCs), in diverse aqueous environments. EDCs have been also analysed by Gu *et al.* with a multichannel two-stage mini-bioreactor system using a genetically engineered bioluminescent bacteria. The toxicity of various samples spiked with known endocrine-disrupting chemicals, and phenol was investigated.

Endocrine Effect Biosensors

Nowadays, there is an increasing concern regarding many environmental contaminants that produce adverse effects by interfering with endogenous hormone systems, the so-called EDCs. EDCs constitute a class of substances not defined by chemical nature, but by biological effect and thus, taking advantage of this feature, "endocrine effect biosensors" have been developed. Steroid hormones induce different effects in mammalian cells after binding to specific intercellular receptors, which are ligand-dependent transcription factors. Many endocrine disruptors are also believed to bind to the estrogen receptor (ER) as agonists or antagonists. Thus, the binding ability of the chemicals towards the ER would be a crucial factor for screening or testing their potential environmental toxicity. Based on estrogen receptors, several biosensors have been developed which provide significant and useful information about estrogenic potency of the sample. The advantage of receptor assays is that they are quite simple to perform and allow the identification of all endocrine disruptors that act through the corresponding receptor.

These biosensors constitute a similar approach as that described above for toxicity biosensors, where effects as a general parameter and not a specific substance are monitored. By employing the commonly utilised human estrogen receptor, the SPR biosensor BIAcore has been applied in the determination of estrogens and xenoestrogens and in binding studies of target compounds. In addition to optical biosensors, electrochemical and piezoelectric biosensors have been developed also based onestrogen receptors.

Biocides

The extensive use of pesticides for agricultural purposes is the cause of their widespread presence in natural waters. Concerns about their toxicity and persistence in the environment has led the European Community to set limits on the concentration of pesticides in different environmental waters: The directive 98/83/EC on the quality of water for human consumption has set a limit of 0.1 μg/l for individual pesticides and of 0.5 μg/l for total pesticides. Although conventional techniques such HPLC/MS and GC/MS gives satisfactory analytical results for pesticide determination, new assays and sensors for cheaper and faster on-site analysis are being developed. Enzymatic sensors, based on the inhibition of a selected enzyme, are the most extended biosensors used for the determination of these compounds.

Based on the inhibition of acetyl cholinesterase (AChE) and colin oxidase, various biosensors have been developed for the detection of organophosphorous and carbamate pesticides such as those described by Choi *et al.*, Vangelis *et al.*, and Andres *et al.* Although sensitive, biosensors based on AchE inhibition are not selective (since the AchE is inhibited by neurotoxins, which include organophosphorous pesticides (OPs), carbamate pesticides, and many other compounds) and cannot, therefore, be used for quantitation of either an individual or a class of pesticides. One approach to solve the lack of specificity of AchE involves the genetic engineering of cholinesterase enzyme to obtain new specific enzymes for desired analytes or families. Different expression systems for the production of recombinant AChEs for biosensor applications were reviewed by Schulze *et al.* The organophosphorous hydrolase (OPH), on the other hand, is able to hydrolyse a number of OP pesticides such as paraoxon and parathion, and chemical warfare agents such as sarin and soman. Hydrolysis of these OP pesticides generates *p*-nitrophenol, which is an electroactive and chromophoric product.

Thus, OPH could be combined with an optical transducer to measure the absorbance of *p*-nitrophenol or with an amperometric transducer to monitor the oxidation or reduction current of this product. With a different approach, the capability of various pesticides such as cyanide, diethyldithiocarbamates, and hydrazines to inhibit the enzyme tyrosinase was also reported. Similarly, diazinon and dichlorvos were detected at limits around 5 μM and 75 μM, respectively, using a tyrosinase-based oxygen sensor. Dithiocarbamate fungicides have been measured by their ability to inhibit the enzyme aldehyde dehydrogenase (AlDH).

Particularly, the detectability of the pesticide Maneb could be improved to a level of 1.5 μg/l, by using a bienzymatic system based on the combination of AlDH and diaphorase. Photosynthesis inhibition is an interesting indicator that rapidly reflects the toxic effect of certain pollutants. Taking advantage of this feature, some biosensors based on Photosystem II (PSII) have been reported to be able to detect herbicides in the environment. About 30 per cent of herbicides, including phenylurea, triazine, and phenolic herbicides, inhibit photosynthetic electron flow by blocking the PSII quinone-binding site and thus modify chlorophyll fluorescence. An amperometric biosensor developed by Koblizek *et al.* exhibited, for example, selective sensitivity to phenylurea and triazine herbicides, whereas phenolic herbicides were not registered. Heavy metals were also able to inhibit the activity of the PSII biosensor, but their effect is usually found at much higher concentrations than those typical for herbicides. Another PSII-based biosensor allowed the detection of herbicides such as atrazine, simazine, isoproturon, and diuron at sub-µg/l concentration levels. In a different approach, biosensors based on immunological assays have been developed. Wilmer *et al.*, determined 2,4-dichlorophenoxyacetic (2,4-D) acid in water by an amperometric immunosensor.

In this immunosensor, the enzyme alkaline phosphatase (AP) catalyses the conversion of PAPP to PAP allowing a limit of detection of 0.1 µg/l. Based on the evanescent wave (EW) transducing principle, atrazine was detected at concentrations around 0.1 µg/l and cyclodiene insecticides in the µg/l range. Mallat *et al.* applied the "River Analyser" (RIANA) immunosensor in the determination of pesticides such as atrazine, simazine, isoproturon, 2,4-D, alachlor, and paraquat in natural waters. Studies focused on the evaluation of matrix effects, interferences due to the presence of cross-reactant substances and on the validation of the sensor.

The system was used for monitoring of atrazine, simazine, and alachlor in different Spanish and Portuguese regions. Based also on SPR principle, another immunosensor was used to measure triazine pesticides achieving a detection limit for simazine of 0.2 µg/l. Analyses carried out on surface and groundwater samples showed a good correlation with parallel chromatographic results. An immunosensor for the determination of Irgarol was applied to the direct analysis of natural waters and water extracts.

It allowed the analysis of samples with an organic solvent content of up to 50 or 25 per cent of methanol and acetonitrile, respectively. In spite of the lack of specificity and the interferences observed in liquid media, some applications have also been reported on the use of piezoelectric immunosensors for the determination of pesticides such as atrazine, 2,4-D, and parathion. Recently, a label-free direct piezoelectric immunosensor built on a flow-through cell was used for the determination of 2,4-D in water with a limit of detection around 0.2 µg/l.

Hormones

Endogenous hormones of human or animal origin have been reaching the environment for thousands of years, even though to an increasing extent they are due to growing population and more intensive farming. Besides endogenous hormones, exogenous sex steroids used as growth promoters in several countries have become a matter of concern not only because of the residues potentially found in meat, but also because environmental soil and water samples may be contaminated as result of the animal excreta. These residues may have endocrine-disrupting activity in aquatic fauna or even terrestrial.

Although very low concentrations (ng/l range) of hormones such as estradiol, estrone, and ethynilestradiol have been found in water, their widespread use and their capability to induce responses in fish at concentrations as low as ng/l or even pg/l level, have alerted scientists to the potential dangerous consequences of their presence in the aquatic environment. As an example, estrone, along with other organic pollutants (atrazine and isoproturon), was determined by Rodriguez-Mozaz *et al.*, with an optical immunosensor in real water samples. The European Union (EU) is funding the development of new biosensing strategies for the control of hormone residues in an effort to improve food quality controls and to protect public health against the adverse effect of these substances.

As a model of example, in the framework of the EU project "Development of single and multianalyte affinity sensors for rapid detection of androgen residues in live an postmortem animals" (RADAR), a group of researchers are developing single and multi-analyte affinity sensors and receptor-based sensors for the rapid detection of androgens such as testosterone and metabolites.

PCBs

Polychlorinated biphenyls (PCBs) are ubiquitous environmental pollutants widely used as industrial chemicals, particularly as dielectric fluids in electrical transformers and capacitors. The high toxicity of some PCB congeners represents a risk for public health as these compounds are still present in the environment, even though the production of PCBs has been banned in several countries many years ago. Different biosensor configurations have been designed to determine PCBs in the environment, and these include the DNA biosensor with chronopotentiometric detection, and various immunosensors with fluorescence, SPR, and electrochemical detection principles.

Dioxins

Apart from PCBs, other polychlorinated compounds of environmental concern are the dioxins, which are released as by-products in a number of chemical processes involving chlorine. Thus, processes such as the production of some pesticides, the manufacture of PVC plastics, the chlorine bleaching of pulp and paper and waste incineration generate dioxins. They are considered carcinogenic and are a potential threat to human health, and more recently, they have been included in lists of potential EDCs. Conventional dioxins analysis requires laborious multistep clean-up procedures that increase the cost of each analysis. A significant number of immunoassays for dioxins have been developed in an effort to provide simplified and routine analysis. The SPR biosensor developed by Shimomura *et al.* for the determination of PCB (mentioned above) was also employed in the determination of the dioxin 2,3,7,8-TCDD. Similarly, another biosensor for detection of dioxin-like chemicals (polyhalogenated dioxins, furans, and biphenyls) based on a recombinant mouse hepatoma cell line was characterised and optimised by Pasini *et al.*

Phenols

Phenolic compounds that appear in the environment originate from the paper and pulp industry and from the production of drugs, dyes, and anti-oxidants. Phenolic compounds, and especially chlorophenols, are important environmental pollutants because of their high toxicity and possible accumulation in the environment. They are also considered as precursors of the dioxins. Parellada *et al.* developed an amperometric biosensor, with tyrosinase (a polyphenol oxidase with a relatively wide selectivity for phenolic compounds) immobilised in a higrogel on a graphite electrode, which correlated satisfactorily with the official method for the determination of the phenol index in environmental samples.

Chlorophenols have been also detected with a flow-injection chemiluminescence fibre optic biosensor, exploiting the ability of certain substituted phenols to enhance the chemiluminescence reaction of luminol, catalysed by horseradish peroxidases. Finally, a Lux-based biosensor was used to assess the toxicity of a paper mill sludge being some metals (Cd and Cu) and pentachloro-phenol (PCP).

Surfactants

Detergent products use surfactants as the basic "active" component. The anionic surfactants are the most widely used, while the cationic surfactants represent only 5 per cent of the total. An amperometric biosensor for detection of anionic surfactants was constructed with *Pseudomonas rathonis T* (bearing a plasmid for surfactant degradation) as a biological element. Oxygen consumption acted as an indicator of cell metabolism and thus of the surfactant content of the sample. The limit of detection achieved for sodium dodecyl sulphate (SDS) was within 0.25-0.75 mg/l. Taranova *et al.* also studied the sensitivity and selectivity of biosensors based on bacterial strain *Pseudomonas* and *Achromobacter* and their ability to degrade the anionic surfactants. The degradation of surfactants by the bacteria caused a decrease in dissolved oxygen and a change in the oxygen electrode current. The microbial biosensor enabled detection of surfactants with high selectivity, sensitivity, and reproducibility. The lower limit reached for SDS was near 0.25 μg/l. Since the 1960s, a new formula with more biodegradable surfactants such as alkylbenzene sulphonates (ASs) was introduced. Linear alkylbenzene sulphonates (LASs) are abundantly used in the formulation of household detergents. Despite the efficient removal of LASs in wastewater treatment plants, residues of LASs are still present in surface waters at the low μg/l range. Even though LASs are not severely toxic, they contribute to the permeation of other pollutants into aquatic animals. A combination of two whole-cell biosensors was applied to river water samples for the determination of anionic surfactants. The first biosensor was based on the detection of the dissolved oxygen consumed in the degradation of LASs by immobilised LAS-degrading bacteria. On the contrary, the other sensor, which used *T. cutaneum* yeast, did not respond to LASs. Values for LAS concentrations derived from the combined use of both biosensors correlated well with values determined using conventional methods.

Alkylphenol ethoxylates (APEs) belong to the group of non-ionic surfactants, whose detection have gained more importance due to their endocrine-disrupting properties. APEs are used on a variety of industrial applications, such as manufacturing of pulp and paper, metals, textiles, paints, resins, adhesives, latex, rubber, and plastics. The APEs have been shown to be estrogenic both in vivo and in vitro. In wastewater treatment processes and in the environment, APEs degrade to alkylphenols (APs), which tend to be more toxic and show greater estrogenic activity. Rose *et al.* described the development of a capillary-based immunoassay (CIA) for APEs and APs, utilising glucose dehydrogenase (DH) as label.

Alkanes, aromatic compounds, and polycyclic aromatic hydrocarbons (PAHs) Contamination of soils and surface and groundwater supplies with petroleum products is a serious environmental problem. Of particular concern for drinking water quality are water-soluble aromatic components (e.g., benzene, toluene, ethylbenzene, and xylenes) of petroleum products. Although many of these contaminants are readily biodegradable, they often persist in the environment. A green fluorescent protein-based *Pseudomonas fluorescens* strain biosensor was constructed and characterised for its potential to measure benzene, toluene, ethylbenzene, and related compounds in aqueous solutions. The biosensor is based on a plasmid carrying the toluene-benzene transcriptional activator. Another microbial whole-cell biosensor, using *E. coli* with the promoter luciferase luxAB gene, was developed for the determination of water-dissolved linear alkanes by luminescence. The biosensor was used to detect the bioavailable concentration of alkanes in heating oil-contaminated groundwater samples.

PAHs are carcinogenic compounds generally formed during incomplete combustion or pyrolysis of organic matter containing carbon and hydrogen. They are very abundant, ubiquitous, and recognised carcinogenic compounds. Amperometric biosensors for naphthalene found in contaminated soils, were constructed using *Sphingomonas yanoikuyae* B1. For benzo(a)pyrene (BaP) and related adducts, a fibre optic fluoroimmunosensor has also been developed in which high sensitivity is achieved by laser excitation and optical detection. The laser radiation reaches the sensor probe and excites the BaP bound to the fibre optic probe.

Antibiotics

Medical substances have been released into the environment with very little attention until recently. The presence of antibiotics in the environment is worrying since they promote antibiotic resistance. The increasing use of antibiotics for therapeutic purposes or as growth promoters in dairy cattle and as feed additives in fish farms or in livestock during the last five decades has caused a genetic selection of more harmful bacteria, which is a matter of great concern. The widespread administration of antibiotics raises significant food safety issues since antibiotic resistance can be transferred to humans on ingestion of affected meat and milk products. Therefore, most of the biosensors developed are aimed at determining them in biological or food samples. For example, a commercial biosensor Biacore 3000 was used to study the cross-reactivity between two sulphonamides (a group of antibiotics): sulphamethazine and furosemide. Sulphonamides sometimes cause allergic reactions, whereas their effect in the human inmunosystem is of high interest for their therapeutical application. Sulphamezathine has been also determined with an optical immunosensor by Akkoyun *et al.* in animal urine. Hansen and Sorensen presented three different reporter gene systems from *V. fischeri*, *E. coli*, and *Aequoreavictoria* all combined with a tetracycline inducible promoter in the development of three corresponding whole-cell biosensors. They respond to low levels of tetracyclines by producing galactosidase, light or green fluorescent protein, respectively. In the field of food monitoring, different biosensors were able to determine penicillin G or tetracyclines, both in milk. More references of biosensors for antibiotic determination can be found in a review by Patel.

Toxins

Toxins are a very heterogeneous group capable of affecting different biochemical processes including membrane function, ion transport, transmitter release, and DNA and protein synthesis. In many cases, specific details of the site and mode of action of a toxin at the molecular level are not known. A number of attempts have been made to detect toxins in environmental and clinical samples using receptor sensors. The induction of reporter genes by a promoter, which responds to a wide variety of toxic compounds, can be used to produce biosensors useful as first indicators of the presence of pollutants in the environment. When the cells are exposed to toxic substances, there can be either a reduction (negative signal) or an activation (positive) in reporter-protein production. Toxin-sensitive cells that express a reporter gene can be used, therefore, as a biosensor for the non-specific detection of toxics, but not for a specific toxin.

A great number of specific sensors for bacterial toxins and mycotoxins have been developed for food and environmental control. Thus, an integrated optical sensor has been reported for the analysis of aflatoxin B in corn. A light-addressable potentiometric immunosensor based on the commercial device (Threshold) for the analysis of saxitoxin and ricin has also been described. An impedance- based immunosensor has been prepared by using an ultrathin platinum film with an immobilised layer of antibodies against the staphylococcal enterotoxin B. Various evanescent wave immunosensors have also been reported to be capable of detecting botulin with very low limits of detection.

A rapid and sensitive immunosensor for the detection of the Clostridium botulinum toxin A has also been developed. This fibre optic-based biosensor utilises the evanescent wave of a tapered optical fibre where antibodies antitoxin A have been covalently immobilised at the distal end. The toxin could be detected by means of a rhodamine label, within a minute at concentrations as low as 5 ng/ml. The reaction was highly specific, and no response was observed against tetanus toxin. With a similar configuration, it has been described for the detection of the cholera toxin. A portable fibre optic biosensor for quantification of the staphylococcal enterotoxin B in diverse media has also been described.

Microorganisms

Bacteria, viruses, and other microorganisms are found widely in polluted, untreated, and treated waters, which implies a worldwide public health problem. In the case of untreated water systems, faecal coliforms may enter rivers and streams through agricultural run-off and non-point source pollution and via domestic wastewater. Pathogenic compounds may reach humans by various routes, such as the use of these waters for recreation or sports, for the irrigation of fruit and vegetables and as drinking water.

Therefore, surface waters may play an important role in the transmission of pathogens. Suitable monitoring of the water supply for the presence of pathogens can assist in preventing disease from these sources. Current methods for detection of pathogenic viruses, bacteria, protozoa, and helminthes tend to be inaccurate, time-consuming, and expensive. As a result, indicator bacteria such as *Salmonella*, "total faecal coliforms", and "total faecal streptococci" are commonly used to determine the risk of faecal contamination and the possible presence of pathogens in water and wastewaters. Other emerging pathogenic microorganisms, *Campylobacter*, *Aeromonas*, and *Yersinia genera* and in recent years, also bacteriophages, have been proposed by several groups as indicators for the microbial quality of water since they are more suitable to signal viral contamination of water sources. On the other hand, monitoring of microorganisms is not only of environmental concern; from a military point of view, there are a number of pathogenic bacteria that can be considered possible biological warfare agents.

Conventional analytical methods for microorganisms are based on colony-forming unit (CFU) count and require selective culture, biochemical, and serological characterisation. In most cases, they require also some enrichment steps. These methods are sensitive and selective, but time-consuming: days to weeks are needed to get a result. The increasing public concern over environment safety has led to a search for technologies capable of rapidly identifying contamination problems at source.

Biosensors capable of detecting an organism quickly will be important in the environmental monitoring of pathogens in real time. The microbial content of a sample can be determined by monitoring microbial metabolism. The transducer can either detect consumption of oxygen or the appearance/disappearance of an electrochemically active metabolite. However, proliferation of nucleic acid and immuno-based detection technologies has provided sensitive and specific detection systems for pathogenic bacteria and viruses. DNA detection may be more specific than immunologically based detection, and the sensitivity can improve thanks to its combination with polymerase chain reaction (PCR) methods.

Gene probes are already finding application in detection of disease-causing microorganisms in water supplies, food, or in plants, animal or human tissues. Immunological detection, on the other hand, is faster and more robust than DNA detection. Moreover, it has the ability to detect not only contaminating organisms, but also their biotoxins. Koubaka *et al.* reported detection of *Salmonella enteriditis* and *Listeria monocytogenes* in real time using an SPR sensor based on antibodies immobilised on the gold sensor surface. *Salmonella* and *Listeria* were detected by the sensor at concentrations down to 106 cell/ml. Recently, a number of piezoelectric biosensors formats have been developed for the detection of several microbial contaminants. *Salmonella typhimurium* detection in liquid samples by an immunosensor based on the acoustic wave principle was reported by Pathirana *et al.* Limits of detection were around 100 cells/ml in less than 2 min.

In another example of piezoelectric biosensors, 23-mer biotinylated probes sequence, specific to *Aeromonas hydrophila*, were immobilised onto the surface of a streptavidin-coated gold surface of a quartz crystal. This sensor was capable of detecting a PCR product amplified from a specific gene of *A. hydrophila* and of distinguishing between samples that contained the gene and samples that did not. Ercole *et al.* described a biosensor for the determination of *E. coli* in water samples by an immunochemical potentiometric alternating biosensor. The monitored change in the redox potential was due to the production of NH3 by a urease-*E. coli* antibody conjugate linked with the *E. coli* cells present in the water.

Hasebe *et al.* described an amperometric tyrosinase-based biosensor for the detection of *E. coli* in wastewater. The detection was based on tyrosinase-catalysed oxidation of polyphenolic compounds, which are produced microbiologically from salicylic acid, and the subsequent signal amplification. The sensor was capable of detecting 103-104 cells/ml after an enrichment step.

The commercialisation of current research in biosensor technology will provide consumers with real-time biosensors capable of maintaining sensitivity better than 100 CFU/ml.

Industries such as Caliper Technologies, Cepheid, Nanogen, ACLARA BioSciences, MICROGEN Systems, and Lawrence Livermore Laboratories are developing microfabricated systems for detection and identification of specific microbial agents. Advances in antibody production and the recent emergence of phage-displayed peptide biosensors offer increased possibilities for the rapid detection of pathogens, but still require time-consuming pre-enrichment in order to detect low numbers of pathogens in water.

Metals

The determination of traces of heavy metals such as Cu, Cd, Hg, and Zn in the environment is very important because of their high toxicity, their increasing environmental levels (due to their use in industrial processes), and because metals can bioaccumulate in living organisms, especially in marine organisms. Metals are usually determined after digestion with strong acids. Common analytical techniques used are ion chromatography, inductively coupled plasma, and polarography. Heavy metals can also be determined with ion-selective electrodes. However, these methods are not able to distinguish between available (potentially hazardous) and non-available (potentially non-hazardous) fractions of metals to biological systems. One advantage of the whole cell-sensors is their ability to react only to the available fraction of metal ions. Recent progress has been made in the development of biosensors relying on intact bacterial cells to monitor toxic metals. Both non-specific and specific biosensors that utilise intact cells have been developed for this purpose. Non-specific microbial biosensors, which have been used for several years, measure only general toxicity, such as the alreadymentioned Cellsense. On the other hand, heavy metals are well known to inhibit the activity of enzymes, and application of this phenomenon to the determination of these hazardous toxic elements offers several advantages such as simplicity and sensitivity. Besides, in many cases, the inhibition effect is related to its biological toxicity. Durrieu and Tran-Minh described a biosensor for the determination of heavy metals based on inhibition of the AP present on the external membrane of *Chlorella vulgaris* microalgae. The microalgae cells were immobilised on removable membranes placed in front of the tip of an optical fibre. Krawczynski vel Krawczyk *et al.* studied the inhibition effect of mercury and other heavy metal ions on urea hydrolysis catalysed by urease immobilised in a polyvinyl chloride (PVC) layer at the surface of a pH electrode in the form of potentiometric biosensor. Urease is the most frequently applied enzyme for inhibition determination of mercury, as it is relatively cheap and easily available. Because the method was not specific, it was applied for the determination of the total inhibition effect caused by heavy metal ions in water samples. The application of another urease-based optical sensor for monitoring low levels of mercury has also been reported by Yerian and Ruzicka.

Urease immobilised into Nafion film on the surface of an ion-sensitive field transistor (ISFET) was also used for heavy metal measurement by Volotovsky *et al.* In this approach, enzyme immobilisation into a negatively charged polymer seemed to cause an increase in the inhibition effect of metals due to cation accumulation in the polymeric matrix. A combination of specific additives and selective rewashing techniques to make the urease-based biosensor sensitive only to mercury ions was proposed.

For the determination of a specific metal, recombinant bacterial sensors have been constructed and used. Specific biosensors, based on inducible promoters fused to reporter genes (e.g., those that code for bioluminescence proteins, such as luciferase), are more sensitive than both chemical analysis methods and non-specific toxicity biosensors. The light emitted can be detected by photometers, illuminometers, and charge-coupled devices. Recombinant luminescent bacterial sensors were used by Ivask *et al.* for the determination of the bioavailable fraction of cadmium, zinc, mercury, and chromium in soil. In this work, two bacterial recombinant heavy metal sensors were constructed based on two different receptor-reporter systems: one was inducible by Zn^{2+}, Cd^{2+}, and Hg^{2+}, and the other by Cr(VI) and Cr(III). The bacterial sensors used were not perfectly specific to one heavy metal, but responded to some "nontarget" metals as well. In another example, the mer-lux gene fusion in *E. coli* was used to estimate bioavailable mercury in soil. The bioavailable fraction was defined here as being part of the water-leachable fraction. The mer promoter was activated when Hg(II), present in the cytoplasm of the biosensor bacterium, binds to MerR, resulting in transcription of the lux genes and subsequent light emission. The luminescence-based bacterial sensor strains *Pseudomonas fluorescens* OS8 (pTPT11) and *Pseudomonas fluorescens* OS8 have also been used for mercury and arsenite detection, respectively, in soil extracts. Soil samples were extracted with water, ammonium acetate, hydrogen peroxide, and nitric acid, and the results obtained were compared with those attained with traditional methods.

Other biosensors have been designed, based on bioengineered proteins. In these cases, the biosensor monitors conformational changes caused by the binding of the metal ion to the engineered protein. Bontidean *et al.* used mercuric ion-binding regulatory proteins as the biological part of the biosensor, MerR. The conformational change resulting from the binding of the metal ion to the protein caused a change in the capacitance, which was proportional to the concentration of the metal ions determined.

Inorganic Phosphate

Inorganic phosphate found in surface waters is used as a measure of the degree of eutrophism. Traditional methods for its determination are chromatography, volumetric titration, or spectrophotometry. Therefore, the development of simple and fast biosensors represents an interesting alternative to them. Various enzymatic phosphate biosensors for phosphate determination have appeared in the literature in recent years. Parellada *et al.* described a configuration based on the sequential action of three enzymes that opens up a way to the construction of reagentless enzymatic phosphate sensors.

Nitrate

The increasing nitrate levels found in ground and surface waters are of concern because they can harm the water environment. In line with this, urban wastewater treatment regulations aim to reduce pollution, including nitrate pollution, from sewage treatment works and industry. A biosensor containing immobilised denitrifying bacteria was applied for the determination of NO_3^- in tap water. Through the reduction—in a reaction chamber, N_2O was formed and determined by a N_2O microelectrode, which was the sensing element of the biosensor. A microscale biosensor for nitrate/nitrite determination was used for in-site monitoring in an activated sludge plant. The biosensor was based on the diffusion of nitrate/nitrite through a tip membrane into a dense mass of bacteria converting the ions into nitrous oxide with subsequent electrochemical detection.

Biochemical Oxygen Demand

Biochemical oxygen demand (BOD or BOD5) is defined as the oxygen required to neutralise organic wastes over 5 days at 20°C, and is a parameter widely used to indicate the amount of biodegradable organic material in water. The conventional BOD test has certain benefits such as being a universal method of measuring most wastewater samples, and, furthermore, no expensive equipment is needed.

It has, however, the limitation of being time-consuming, and consequently it is not suitable for online process monitoring. Thus, it is necessary to develop an alternative method that could circumvent the weakness of the conventional BOD test. Fast determination of BOD could be achieved by biosensor-based methods. A common feature of these sensors is that they consist of a microbial film that can biooxidise the organic substrate to be quantified, sandwiched between a porous cellulose membrane and a gas-permeable membrane as the biological recognition element. The response is usually a change in concentration of dissolved oxygen or other phenomena such as light emission. Most BOD sensors rely on measuring the bacterial respiration rate in close proximity to a transducer, commonly Clarktype (an amperometric sensor for measuring dissolved oxygen developed by Clark in 1956). Some BOD sensors have been developed and marketed by various manufactures in both biofilm and bioreactor-type configurations.

Most commercially available BOD sensors are flow-type systems that can be more easily automated, but generally require high maintenance to prevent fouling and clogging. Instrument information about many BOD commercial biosensors was provided by Liu and Mattiason. Despite the good agreement between biosensor results and conventional BOD analysis, and despite the short response time of biosensors, current BOD biosensor systems still present a series of limitations that restrict their industrial applicatios: the lack of standardisation and legislation in most countries, complicated maintenance requirements, and insufficient resistance to various toxic compounds such as heavy metal ions, CN^- and phenol in the wastewater. It is possible to eliminate the toxic effects of heavy metal ions by using a chelating agent that complex the ions, e.g., ethylene diamine tetra-acetate (EDTA) and sodium diethyl dithiocarbamate (DDTC). Prevèntion of contamination by other microbes is also important for a reliable biofilm-type BOD sensor.

Many BOD biosensors have been developed for the determination of high BOD values in industrial wastewater and not adapted to the measurement of low BOD values. An optical fibre biosensor was developed for the evaluation of low BOD values in river waters by Chee *et al.* The immobilised *Pseudomonas putida* bacterium membrane was placed on the top of an optode, which was linked to a photodiode that detected fluorescence signal. The response time was 15 min for chloride up to 1000 mg/l.

Bioremediation

Bioremediation is an application of the microbial capacity to transform complex organic molecules into simpler inorganic constituents. Parameters such as nutrient availability, metal ions, pH, dissolved oxygen, and temperature, influence the growth of bacteria. Biosensors that can monitor these parameters will help to better control the bioremediation process. Different molecular biosensors implemented to monitor these parameters were reviewed by Purohit. These biosensors use the luciferase expression system. The biological component in this molecular biosensor is a recombinant plasmid. It has a specific promoter, whose expression is sensitive to a target molecule.

Biosensors as Detectors of Separation Methods

Although chemical separations can be highly efficient, some mixtures are too complicated to be separated by chromatography. At the same time, biosensors suffer often from lack of selectivity in the determination of certain compounds. The combination of these two methods can enable identification of molecules that are not detectable separately by either method. Components are thus identified by both functional recognition and separation retention time. A review by Fishman *et al.* presented examples and principles of combining chemical separation with biosensor detection using living systems, whole cells, membrane receptors, enzymes, and immunosensors.

Commercial Biosensors

Despite the high number of biosensors under development and also the amount of research literature on this area, few practical systems are currently enjoying market acceptance. The first successful commercial biosensor was the "glucose pen", launched by Exactech in 1987. Nowadays, 90 per cent of sales come from glucose-detecting biosensors for medical applications. However, in other areas such as in food, agriculture, military, veterinary, and the environment, there is a potential market still to be established.

Many of the instrumentations developed for the medical diagnostics market could be adapted for environmental market, for example. Even though commercial returns from environmental biosensors are substantially less than from medical diagnostics, public concern and government funding has generated a major research effort aimed at the application of biosensors to the measurement of pollutants and other environmental hazards.

SPR biosensors constitute the most successful type in the commercial instruments for environmental monitoring. The pioneers of SPR-based biosensing were Pharmacia Biosensor AB, now BIACORE AB (Uppsala, Sweden), who launched the original BIAcore system in 1990. The company now has a large range of biosensors, which includes several generations of the original BIACORE (series 1000, 2000, and 3000) as well as other configuration systems offering varying degrees of automation and parameter specifications. The BIACORE 3000 biosensor, which is based on SPR principle, monitors biomolecular interactions as they proceed over time.

The advantage of the method is that there is no need for labelling the interactants. Approximately 90 per cent of the 1998 and 1999 commercial biosensor publications cite the use of BIACORE instruments. Other SPR commercial biosensors are provided by Windsor Scientific, Ltd. (Berks, UK), which markets the IBIS system; and by Nippon Laser and Electronics Labs (Hokkaido, Japan), which markets CELLIA sensor. CELLIA systems can be configured for either whole cells or macromolecules.

Texas Instruments (Dallas, TX) integrates SPR detector Spreeta that can be configured for industrial, environmental, and biological applications. BioTul AG (Munich, Germany) launched its Plasmon SPR instrument in August 1999. On the other hand, Affinity Sensors (Franklin, MA) manufactures the IAsys line of instruments, which use evanescent wave technology. In contrast, REMEDIOS, a whole cell-based biosensor, can be employed for diagnosis of contaminated lands or soils. It detects the levels of any toxicity that affects the metabolic activity of the biosensor organisms. Under normal conditions, the biosensor gives out visible light. The light output is directly proportional to the metabolic activity. If the sample (water, soil, sediment, or sludge) has any available toxicity, the bioluminescence will decrease in direct proportion to the level of toxicity. Biosensor organisms are selected as representative of bacterial strains found in the environment as well as those that are involved in bioremediation processes.

Future Perspectives

In spite of the past and current large amount of research in biosensor development, there is still a challenge to create improved and more reliable devices.

Nature is full of intricate nanosystems, far more advanced than all of the man-made systems. A significant effort is being made nowadays to either mimic or use such systems in the development of new applications and devices such as sensors and biosensors. In order to improve biosensing interfaces, design and analysis on the molecular level have been proposed. All are addressing biointerfaces in molecular dimensions and thus can be summarised by the term "molecular nanotechnology". Therefore, there is a trend towards the combination of physics and biology in the creation of new nanostructures.

Nanotechnology comprises a group of emerging techniques from physics, chemistry, biology, engineering, and microelectronics that are capable of manipulating matter at nanoscale. This novel technology bridges the gap between materials science, coming from the micrometer range, and biochemistry, where individual molecules are of major interest.

Inspired by nature, molecular self-assembly has been proposed for the synthesis of nanostructures capable of performing unique functions. Self-assembly is the formation of organised, patterned structures without external direction. Biomaterials, such as proteins, lipids, and nucleic acids, can self-assemble. Such nanostructures are applied for the development of amperometric immunosensors.

Tiefenauer *et al.* have developed a nanostructured gold film electrode with openings where the biotinilated antigen is bound to the capture molecule streptavidin (covalently immobilised on silica). The close contact of the recognition and redox centres to the electrodes is designed to facilitate electron transfer. Several research groups have begun to explore alternative strategies for the development of optical SPR biosensors based on the extraordinary optical properties of metal nanoparticles.

Metal nanoparticles that differ in size and composition can be designed as well, to scatter light of different wavelengths according to their distinct SPR. Cao *et al.* showed how oligonucleotides and presumably other biomolecules (e.g., proteins) can be used to modify the surfaces of such particles, thereby imparting useful biorecognition properties to them. For example, Lazarides *et al.* have developed gold colloidal nanoparticle aggregates that are linked by short pieces of DNA. These materials exhibit a colour change from red to blue after DNA hybridisation. This colour change is due to electromagnetic coupling between the gold non-spherical nanoparticles.

Lundstrom *et al.* have proposed the use of pigment-containing cells (present in the skin of certain fishes and frogs) as biosensors. The cell membrane receptors, the G-protein coupled receptors, control the pigment particles aggregation, which is simple to measure. This natural nanosystem shows large promise for new biosensing and bionanalytical systems since, through several biochemical modifications, G-protein coupled receptors of almost any kind can be provided.

Multianalyte Determination

Development of sensors capable of determining several analytes simultaneously can represent an interesting tool in environmental monitoring and screening. This configuration allows the reduction in time and sample volume and other reagents required. Different technologies have been developed in recent years to produce multianalyte sensors. The development of large-scale biosensor arrays composed of highly miniaturised signal transducer elements, for example, enables the real-time parallel monitoring of multiple species and is an important driving force in biosensor research. In clinical diagnosis, simultaneous determination of multiple analytes has been a great challenge.

Multichannel biosensors are required for direct detection in high-throughput screening systems in the search for new pharmaceuticals. An example of multichannel performance is the biosensor proposed by Berger *et al.* using BIAcore system that demonstrates the feasibility of a multisensing device for monitoring four separate immunoreactions simultaneously in real time.

A planar array immunosensor equipped with a charge-coupled device (CCD) as a detector and a diode laser as light source, has been also developed and applied to either the determination of multiple compounds, such as virus, toxins, and bacterial spores in a single sample analysis or a single analyte in multiple samples simultaneously.

For environmental applications, a multianalyte enzyme biosensor based on a disposable, thickfilm multielectrode was developed for the analyte discrimination of binary mixtures of the pesticides paraoxon and carbofuran. Other studies aimed at the development of multianalyte immunosensors, such as that described by Gonzalez-Martínez *et al.*, which developed a competitive capture assay for carbaryl, atrazine, and irgarol as target compounds. Mastichiadis *et al.* have applied recently an optical capillary immunosensor to the simultaneous determination of the pesticides mesotrione, hexaconazole, paraquat, and diquat, by preparing an ordered array capillary of four distinct analyte bands.

After the optimal characterisation of analytical conditions in deionised water, further experimentation is required in order to apply the biosensor to natural water sample analysis. Barzen *et al.* and Rodriguez-Mozaz *et al.* have described the application of a multianalyte immunosensor based on total internal reflection fluorescence, to the simultaneous determination of different analytes in the same sample. Atrazine, isoproturon, and estrone were successfully determined in natural water samples with this biosensor.

Miniaturisation

In general, there is a growing tendency towards miniaturisation of analytical systems, since it allows the handling of low-volume samples, a reduction in reagent consumption and waste generation, and increases sample throughput. Taking advantage of miniaturisation benefits, sensors and biosensors can become inexpensive and easy-to-handle analytical devices for fast, reliable measurements of chemical species.

Miniaturisation emerges thus as an important aspect of future bioelectronics. Biosensor miniaturisation has a particular significance for medical applications when sometimes an implantable sensor is desired for continuous in vivo monitoring, and for pharmaceutical industries, in the field of high-throughput screening. For environmental monitoring, the size of a sensor is usually not so important, and there is not an obvious need to produce miniaturised systems.

However, the use of a miniaturised flow cell and a microsensor will become a key technology, reducing the amount of used chemicals, which will have an impact not only on the environment, but also on the economy.

There is, in fact, an increase in the demand on environmental analysis and also the amount of chemicals to be determined. The time and expense devoted to the detection of environmental pollutants are becoming impediments in treating a number of samples for many environmental projects. In this sense, microfabricated sensors or systems will become a powerful tool. On the other hand, a small size would be preferred in the design of portable biosensors, especially for on-field screening applications.

Another advantage of miniaturisation at the nanometer-scale level is that it will allow high-density information storage. This miniaturisation trend involves adaptation of microfabrication and nanofabrication techniques. In addition, the practical application of any sensor requires the use of complete analysis systems for sample handling such as pumps, filters, membranes, and sample conditioning.

Such formats may be used in bioanalytical microsystems that comprise another new trend in bioanalysis. In such microfluidic systems, miniaturised biosensor arrays, as well as miniaturised sampling, filtering, and so on, has to be accomplished.

Mass Production

Technologies usually enable not only miniaturisation, but also batch-fabrication, which leads to the production of inexpensive sensors. Automated manufacturing technologies are required in the commercial production of large numbers of inexpensive, reproducible biosensor devices. Mass fabrication methods have enabled biosensors to be produced in large volumes for clinical markets.

Two techniques borrowed from the electronics industry have proved particularly important: screen-printing (ink is pressed through a mask to form a pattern on a ceramic or plastic base) and photolithography (a photoresistant material is exposed to ultraviolet light passed through a mask and then the silicon is chemically etched). Silicon fabrication technology has been preferred as a means of mass production of biosensors. This is an area of intense research activity mainly by diagnostics companies. The use of silicon microfabrication for both electrochemical and optical sensors is expanding, and the capability of on-chip electronic signal amplification and data processing are very attractive.

Bookham Technology (Oxfordshire, UK) is using microelectronics technology based on silicon to produce optical chips for biosensor application. Other techniques such as airbrush and Cabro deposition have been adapted for biosensor fabrication, especially for thick-film biosensors. Thin-film sensors are generally produced by a variety of vapour deposition techniques, electrochemical methods and more recently by the use of Langmuir–Blodgett technology. These have been used in the production of sensors based on SPR.

Laser desorption is particularly promising and offers the ability to "write" proteins to surface with very high resolution (patterning).

Network Systems

The future integration of wireless communications systems will generate huge interest as this will inevitably lead to the emergence of extensive networked multiple autonomous analytical stations in rivers, lakes, wells, or even water treatment plants. These units will provide high-quality information about key chemical parameters that determine the quality of our environment. By means of this network system and by an intelligent remote surveillance, an active response will be possible. To accomplish this, there is a need for robust and unattended working analytical instrumentation. Each unit will be portable, plug-and-playable, user-friendly, and fully automated with little need for maintenance.

The emergence of these compact, self-sustaining, networked instruments will have enormous impact on all field-based environmental measurements including biosensors. Gu *et al.* have recently applied a novel early warning protocol to monitor the toxicity of the effluents of a water treatment plant. A toxicity early warning system is needed to prevent the outflow of water containing dissolved toxic chemicals and is important to ensure compliance with EU directives.

Validation

Despite the practical advantages of biosensors, they must be able to provide results in real time, must be simple to use, portable, and cost-effective. Acceptance of these methods is dependent upon several factors: they should be comparable to conventional analytical systems in terms of reliability, sensitivity, selectivity, specificity, and robustness. Therefore, biosensor measurements have to be verified by comparing them with the results of chemical analysis.

This verification process is known as validation. Method validation is the process of demonstrating that the combined procedures of sample preparation (extraction, clean-up, etc.) and analysis will yield acceptably accurate, precise, and reproducible results for a known analyte in a specific matrix. Thevenot *et al.* recommend several analytical parameters needed to define in order to characterise biosensor performance.

They include four sets of parameters: (i) calibration characteristics: sensitivity, working and linear concentration range, limit of detection (LOD) and of quantification (LOQ); (ii) selectivity and reliability; (iii) steady-state, and transient response times and sample throughput; (iv) reproducibility, stability, and lifetime.

The overall commercial status and general acceptance of the technology will depend on the performance characteristics, sample throughput, associated costs, and acceptance by regulatory authorities, based on independent validation data generated using internationally recognised procedures.

Bioengineering

Biosensor sensitivity and selectivity depend essentially on the properties of the biorecognition elements to be used for analyte binding. Genetic engineering provides an elegant way not only for providing unlimited amounts of biorecognition molecules, but also for the alteration of existing properties and the supplementation with additional functions. The current generation of biosensors is taking advantage of novel biosensing materials based on gene engineering. In the field of biosensors, gene engineering focuses on two main fields: (1) genetically engineered receptor molecules, and (2) genetically transformed cells. Selective receptor units of the sensor are usually derived from subcellular components. The most commonly applied biomolecules are enzymes, antibodies, and receptors. Current research includes combinatorial screening or design of the amino acid sequence of the binding region of antibody.

In this sense, recombinant antibodies provide an emerging strategy in the development of new immunosensors. A recently developed method that generates antibody-like molecules that bind specifically to target molecules is "phage display". Phage display affords a unique way of selecting peptides and proteins with binding affinity similar if not higher than monoclonal antibodies.

These techniques make possible the production of high amounts of affinity peptides or antibodies without the immunisation and sacrifice of laboratory animals. Once isolated, the displayed peptide can be easily examined to determine its sequence and subsequently may be manipulated for improved stability or affinity. In phage display, a library of peptides or larger molecules, including antigen binding fragments or single chain fragment variables, is encoded within a population of bacteriophage, which express them on the surface. These libraries can be applied to select and isolated specific suitable antibodies or peptides for any antigen.

Different authors are now investigating the potential of using phage-displayed peptides as reagents in sensor applications. Benhar *et al.*, for example, presented single-chain phage-displayed antibodies combined with amperometric detection and its application as an immunosensor for the detection of three different analytes. Goldman *et al.* demonstrated the potential of a phage-based biosensor for the determination of a staphylococcal enterotoxin B (SEB). A library of random 12-mers displayed on phage was panned against SEB. Whole phage was labelled with the dye Cy5, allowing fluorescence detection of the assay. Recombinant antibodies derived from antibody libraries with *s*-triazine selectivities were applied and modified by Hock *et al.* for the development of suitable antibodies likely to be used in immunoassay or immunosensors.

Modified cell biosensors are constructed by fusing a reporter gene to a promoter element that is induced by the presence of a target compound. This genetic information, located on a plasmid vector, is then inserted into a bacterial strain, which may be integrated into the chromosome or remain in plasmid form. The engineered genetic fusion is then replicated along with the cell's normal DNA. As reported above, in the case of bioluminescence sensors, the recombinant bacteria employed contains gene fusions between the regulatory region of the mer operon (*merR*) and bacterial luminescence genes, which respond to contaminants. Recombinant luminescent bacterial sensors were used by Ivask *et al.* for the determination of cadmium, zinc, mercury, and chromium.

When Hg(II), present in the cytoplasm of the biosensor bacterium, binds to *merR,* the mer promoter is activated, resulting in transcription of the lux genes and subsequent light emission. Three different whole-cell biosensors for the determination of mercury were made using *E. coli* by fusing the mercury-inducible promoter, *Pmer*, and its regulatory gene *merR,* with three different reporter genes producing light, a-galactosidase or green fluorescent protein, respectively, as response to the presence of low levels of mercury. The same group presented the choice of three different reporter gene systems combined with a tetracycline-inducible promoter in the development of three different biosensors for the analysis of tetracyclines as reported above.

Despite the huge potential of biosensors, and the ever-increasing number of biosensors developed, commercially available biosensors are being applied to a restricted area of the potential market. In general, biosensors for environmental analysis have several limitations: sensitivity, response time, and lifetime, which should be improved for them to become a competitive analytical tool.

The areas of development that are expected to have an impact in biosensor technology are: immobilisation techniques, nanotechnology, miniaturisation, and multisensor array determinations. However, a crucial aspect may be the production of new sensing elements easy to synthesise and with the capability to broaden the spectra of selectivities that can be reached by a biosensor. At present, the preparation and production in large scales of biomolecules such as enzymes or antibodies need an investment of time and knowledge.

Synthetic peptides and MIPs are contemplated as promising alternatives overcoming the above-mentioned limitations. Unfortunately, the affinity accomplished by these synthetic receptors is still several orders of magnitude below that of the antibodies. Improvement in the affinity, specificity, and mass production of the molecular recognition components may ultimately dictate the success or failure of detection technologies. The possibility of tailor binding molecules with predefined properties, such as selectivity, affinity, and stability, is one of the major aims for biotechnology. The development of advanced receptors will allow the analysis of complex real samples and *in situ* measurements resolving the responses from the analyte and from non-specific background effects. Since scientific attention is currently being given to biotechnology, as this review has pointed out, the development of improved molecular recognition elements will be followed by a corresponding enhancement of the biosensor features.

From the above viewpoint, it is clear that the future of biosensors will rely on the success of emerging sophisticated micro and nanotechnologies, biochemistry, chemistry, thin-film physics, and electronics. To reach this goal, an important investment in research, expertise, and the necessary facilities is needed. However, as the world becomes more concerned about the impact that environmental contamination may cause on public health and the ecosystem, the demand for rapid detecting biosensors will only increase. Biosensors still need to achieve the confidence of potential users, having in mind that the commercialisation of new devices will always be the best indicator of the success of a biosensor technology.

The analysis of complex matrices and of analytes difficult to determine by the actual analytical procedures (i.e., highly polar compounds), are progressively being approached by biosensors. However, there is still a lack of alternative biosensing systems for an important bunch of emerging contaminants such as bisphenol A, phtalates, and polybrominated compounds (used as flame retardants), veterinary and human medicines and personal care products (nutraceuticals, synthetic fragances, sun screen agents, etc.).

7

Genetic Exchange in Environment

Reports made over the past few years leave little doubt that horizontal gene transfer among bacteria can and does occur in the environment. The significance of these events in the dissemination of recombinant molecules is not clear and more research is required to fully appreciate the impact of genetic exchange on their stability and survival. Based on the available data, it is fair to say that horizontal gene transfer may play as important a role in transmission of recombinant sequences as does the survival and persistence of genetically engineered microorganisms.

Genetic Exchange in Soil

Degraders and Indigenous Biphenyl Degraders: *Pseudomonas aeruginosa* JB2, a chlorobenzoate degrader, was inoculated into soil having indigenous biphenyl degraders but no identifiable 2-chlorobenzoate (2CBa) or 2,5-dichlorobenzoate (2,5DCBa) degraders. The absence of any indigenous chlorobenzoate degraders was noted by the failure to obtain enrichment cultures with the addition of 2CBa, 3CBa, or 2,5DCBa and by the failure of soil DNA to hybridise to the *tfdC* gene, which encodes ortho fission of chlorocatechols. In contrast, DNA extracted from inoculated soils hybridised to this probe. Bacteria able to utilise both biphenyl and 2CBa as growth substrates were absent in uninoculated soil, but their presence increased with time in the inoculated soils.

This increase was related kinetically to the growth of biphenyl degraders. *Pseudomonas* sp. strain AW, a dominant biphenyl degrader, was selected as a possible parental strain. Eight of nine recombinant strains, chosen at random, had high phenotypic similarity (90 per cent or more) to the inoculant; the other, strain JB2-M, had 78 per cent similarity.

Two hybrid strains, *P. aeruginosa*JB2-3 and *Pseudomonas* sp. JB2-M, were the most effective of all strains, including strain AW, in metabolising polychlorinated biphenyls (Aroclor 1242). Repetitive extragenic palindromic-PCR analysis of putative parental strains JB2 and AW and the two recombinant strains JB2-3 and JB2-M showed similar fragments among the recombinants and JB2 but not AW. These results indicate that the *bph* genes were transferred to the chlorobenzoate-degrading inoculant from indigenous biphenyl degraders.

Xenobiotic compounds which are readily biodegradable frequently require only a single organism for biodegradation, while persistent compounds such as polychlorinated biphenyls (PCBs) require more than one organism. PCBs are cometabolised by biphenyl-utilising bacteria to chlorobenzoates in culture and in the environment. Furukawa and Chakrabarty were the first to suggest combining the genes of chlorobenzoate degraders with those of biphenyl degraders for complete mineralisation of PCBs. Single strains able to utilise chlorobiphenyls have recently been constructed by genetic complementation of the catabolic pathways from parental strains containing the *bph* (biphenyl) and *clc* (chlorocatechol) genes. Presently, no strains have been constructed which can utilise environmentally relevant PCBs.

An alternative strategy to the development of recombinants in the laboratory is to produce them *in situ*. Although biphenyl degraders are common in the environment, chlorobenzoate degraders appear not to be. Thus, inoculation of soil with the "missing genes" may allow a recombinant strain to arise that could mineralise PCBs. Fulthorpe *et al.* showed that genes for 3-chlorobenzoate (3CBa) degradation persisted in an aquatic community as a result of genetic exchange, even though the inoculated 3CBa degrader did not survive.

Altamont soil was amended with 4,000 mg of biphenyl 100 mg and 300 mg of NH_4NO_3 and was periodically sampled for 2CBa degraders and biphenyl degraders according to previously described procedures. Inoculation of soils was carried out with *Pseudomonas aeruginosa* JB2 obtained from late exponential growth in a mineral salts (MS) medium containing 2,5-dichlorobenzoate (2,5DCBa). Washed cell suspensions were prepared in sterile 50 mM phosphate buffer (pH 7.0) at a density of 10^8 cells ml^{-1}, and 18 ml was added to 100 g of soil to give about 50 per cent saturation of the pore space volume. Sterile 50 mM phosphate buffer was added to uninoculated microcosms.

Periodically, a measured amount of soil (about 1 g) was removed for dilution plate counts by using Noble agar (Difco) containing either 2-chlorobenzoate or biphenyl as the sole growth substrate. Biphenyl crystals were added to the bottom lid of the inverted petri plate for growth on MS agar plates. Colonies from 2CBa plates were transferred to biphenyl plates to determine the frequency of isolates able to grow on both substrates. Colony formation was not evident from dilutions greater than 10^{-3} on agar plates containing neither 2CBa nor biphenyl.

Recombinants capable of utilising both 2CBa and biphenyl were first isolated on 2CBa agar from soil at 15 days. Nine such isolates were randomly selected for further analysis. All recombinant strains grew on 2,5DCBa, 3CBa, 2CBa, and biphenyl. All but one of the nine showed strong phenotypic similarity (90 per cent or higher) to *P. aeruginosa* JB2 when compared by Biolog GN microplates (Biolog Inc., Hayward, Calif.). The exception was JB2-M (similarity, 78 per cent). A dominant biphenyl utiliser, unable to metabolise chlorobenzoates, was obtained from dilution plates and identified as *Pseudomonas* sp. Strain AW. Strains JB2, AW, and all nine recombinant strains were catalase and oxidase positive, did not ferment glucose, showed no change in litmus milk, and, except for strain AW, were resistant to kanamycin, chloramphenicol and tetracycline at concentrations of 100 mg. liter^{-1}. All strains except AW and JB2-M were fluorescent on King's A and King's B agar and grew on tryptic soy agar with 1,10-phenanthroline (considered indicative of *P. aeruginosa*).

Recombinant strains were unstable, and retention of the ability to utilise both biphenyl and 2,5DCBa required alternate transfers between both substrates; if cultured on one substrate continuously (either biphenyl or 2,5DCBa), the recombinants lost the other catabolic pathway. However, no plasmid DNA was isolated from the recombinants. Recombinant strains JB2-3 and JB2-M were also found to be unstable when added to Altamont soil in another experiment conducted under similar conditions.

Cultures, extracted and analysed by gas chromatography after 3 weeks, showed marked differences in the ability to cometabolise Aroclor 1242 (100 mg. litre^{-1}) when grown with biphenyl (500 mg. litre^{-1}). Two strains would not grow in the presence of PCB, three strains transformed less than 2 per cent, two strains transformed between 20 and 28 per cent, and the most effective strains, JB2-M and JB2-3, transformed 56 and 37 per cent, respectively, of the PCBs. These levels were significantly greater than that of the indigenous strain AW (27 per cent). Neither strain AW nor the recombinants could grow on Aroclor 1242 as the sole carbon source. The greater cometabolism of PCBs by recombinant strains JB2-M and JB2-3 than by strain AW may be explained by the elimination of inhibitory intermediates produced in PCB transformation. Sondossi *et al.* found that chlorobenzoates could be fortuitously transformed by biphenyl-induced dioxygenases, leading to formation of chlorocatechols and muconic semi-aldehydes, which in turn inhibit 2,3-dihydroxybiphenyl-1,2- dioxygenase.

Strains JB2-M, JB2-3, and AW were investigated in more detail to assess their genetic relatedness. For extraction of DNA, parental and recombinant strains were grown in 50 ml of MS medium containing either biphenyl or 2CBa. Cells were pelleted (6,000 × *g* for 10 min) and resuspended in 8 ml of TE buffer (50 mM Tris, 20 mM EDTA, pH 8.5), and 0.5 ml 20 per cent sodium dodecyl sulphate (SDS) was added and the mixture was heated at 658°C for 15 min. The mixture was cooled to 508°C and incubated for 2 h after the addition of 1 ml of pronase E (5 mg/ml; Sigma) and 0.2 ml of proteinase K (20 mg. litre^{-1}; United States Biochemical Corp., Cleveland, Ohio). Three phenol-chloroform extractions were followed by one chloroformisoamyl alcohol extraction. DNA was collected by ethanol precipitation and digested with *Bam*HI and *Hin*dIII (New England Biolabs, Beverly, Mass.).

Repetitive extragenic palindromic (REP)-PCR determinations were obtained from cell pellets which had been washed once with TEN buffer (50 mM Tris-HCl, 20 mM EDTA, 100 mM NaCl) containing 0.5 per cent sarcosine and then resuspended in 600 ml of TEN buffer containing SDS (1.25 per cent) and proteinase K (0.63 mg. per litre^{-1}) and incubated at 378°C for at least 1 h. Chromosomal DNA was extracted from the lysed cells, and REP-PCR determinations were performed by using 100 ng of template DNA per reaction.

REP-PCR banding patterns indicated that the two recombinant strains JB2-M and JB2-3 were very similar to the inoculant strain JB2. The origin of the indigenous parental strain could not be resolved, as the likely candidate, *Pseudomonas* sp. strain AW, showed no similarity in banding patterns. The dominant band sizes, ranging from 210 to 5,000 bp, were also similar to those of amplified fragments from *P. aeruginosa* PAO1C. The PCR pattern of *Escherichia coli* HB101 was also similar to that reported elsewhere, indicating that similar PCR conditions were maintained throughout the experiments. Our data indicate that the inoculant strain, by virtue of its genetic similarity to the recombinants, would be the recipient of *bph* genes donated from indigenous biphenyl utilisers. However, recombinants were selected on 2CBa agar and then transferred to biphenyl agar. Selection on biphenyl agar may have resulted in recombinants that were more similar to strain AW, or another indigenous biphenyl utiliser, and less similar to strain JB2.

As enrichment culture from uninoculated soil failed to yield any indigenous bacteria able to utilise 2CBa, 2,5DCBa, or 3CBa after 21 days, *P. aeruginosa* JB2 was assumed to be the parental strain contributing the CBa genes. To verify that non-culturable indigenous chlorobenzoate degraders were not present, DNA was extracted from soils inoculated with strain JB2 and from uninoculated soils and probed with the *tfdC* gene, which encodes catechol 1,2-dioxygenase II. DNA was extracted from the Altamont soil and purified by passage through Sephadex columns. Digestion with restriction enzymes was not done because of shearing during extraction.

Hybridisations were performed under high stringency conditions (>70 per cent of base pairings) with random-primed dioxigeninlabeled DNA by using the Genius 1 Non-radioactive DNA. Labelling and Detection kit (Boehringer Mannheim, Indianapolis, Ind.) and were detected by chemiluminescence (LumiPhos 500) with the Biodot SF slot blot system (Bio-Rad, Hercules, Calif.). DNA from strain JB2, from recombinant strains JB2-3 and JB2-M, and from inoculated soil hybridised to the probe, while DNA from both uninoculated soils did not.

Genomic DNA of *E. coli* HB101, supposedly a negative control, has been previously reported to hybridise to a related probe, due to similar conserved sequences. Genetic exchange can be described as a second-order reaction dependent on the concentration (cell g^{-1}) of chlorobenzoate degraders (A) and biphenyl degraders (B) to produce a concentration of cells having both traits (C):

$$dC/dt = KAB$$

where K is the rate constant ($cell^{-1}\ g^{-1}\ day^{-1}$). This model is based on recombinants being formed by genetic exchange without their subsequent growth. Because ecombinant strains JB2-3 and JB2-M were unstable when grown continuously on biphenyl in pure culture and in soil, it is unlikely that they would grow on biphenyl while retaining the ability to utilise 2CBa. As a relatively constant density of chlorobenzoate degraders of about 10^8 cell g^{-1} was observed after 10 days, the increase in recombinants between 10 and 40 days was dependent upon an increase in biphenyl degraders. For example, the rate of genetic exchange would be 100 times greater at 40 days than at 10 days. An approximation of the rate constant can be obtained from the exponential growth phase of biphenyl utilisers:

$$dB/dt = \mu B - KAB$$

where m is the growth rate constant (day21, in common log notation). Dividing both sides by B gives:

$$d\log B/dt = \mu - KA$$

The slope ($d\log B/dt$) of 0.07 day^{-1} is the net difference between the growth rate μ and KA. Although we cannot determine the growth rate, it must be greater than the slope and it is not likely to be greater than what was found (1.1 day^{-1}) in an earlier study with Altamont soil (10). For $A \cong 108$ cell g^{-1} and $0.08 < \mu < 1.1$, the rate constant K is between $1.0 \,.\, 10^{-10}$ and $1.4 \,.\, 10^{-9}$ $cell^{-1}\ day^{-1}$. Thus, the rate of recombination at 40 days would be between $1.0 \,.\, 10^{6}$ and $1.4 \,.\, 10^{7}$ cells $g^{-1}\ day^{-1}$.

Although the dominant biphenyl-degrading strain AW would be stochastically the most likely parental donor, REPPCR analysis showed no evidence of similar conserved sequences. Nevertheless, the amplified sequence that was used may not necessarily be related to the *bph* genes transferred to strain JB2. The presence of non-culturable biphenyl degraders, genetically more similar to strain JB2 and having a higher K value than strain AW, cannot be precluded early in the experiment

(15 days) when they would not be as greatly outnumbered by strain AW than they would be near the end (45 days).

The occurrence of genetic exchange in the environment has been well documented with respect to biodegradation. Fulthorpe and Wyndam observed an increase in cells having homology for the *cba* region of pBR60, which encodes for 3CBa metabolism, despite a decline of the 3CBa-degrading inoculum. That genetic exchange of degradative genes should occur is not surprising as many of them are found on plasmids.

Although we could find no evidence of plasmids in the recombinant strains or in the recipient, integration into the host chromosome may have occurred by a transposon. Transposons have been reported with genes coding for degradation of biphenyl, chlorohydroxyquinone, chlorobenzoate, trichlorobenzene, 3,4-dichlorobenzoate, naphthalene, and for the inoculant.

The recombinant strains isolated were found to be unstable; if cultured continuously on either substrate (i.e., biphenyl or 2CBa), they would lose the ability to grow on the other. Instability of catabolic genes may be caused by plasmid or transposon loss under conditions which do not select for that genetic element. There are examples of unstable chlorobenzoate and biphenyl genes. However, the high rate of genetic exchange, estimated from the kinetic data, may render concern about stability of recombinants moot. While this instability presents difficulties in maintaining stable cultures or inoculants for bioremediation, it may be desirable in minimising perceived risks of persistent recombinants in the environment.

Ice-minus Bacteria

Ice-minus bacteria is a nickname given to a variant of the common bacterium *Pseudomonas syringae* (*P. syringae*). This strain of *P. syringae* lacks the ability to produce a certain surface protein, usually found on wild-type "ice-plus" *P. syringae*. This protein found on the outer bacterial cell wall acts as the nucleating centres for ice crystals. This facilitates ice formation, hence the designation "ice-plus." The ice-minus variant of *P. syringae* is a mutant, lacking the gene responsible for ice-nucleating surface protein production. This lack of surface protein provides a less favourable environment for ice formation. Both strains of *P. syringae* occur naturally, but recombinant DNA technology has allowed for the synthetic removal or alteration of specific genes, enabling the creation of the ice-minus strain.

Production

To systematically create the ice-minus strain of *P. syringae*, its ice-forming gene must be isolated, amplified, deactivated and reintroduced into *P. syringae* bacterium. The following steps are often used to isolate and generate ice-minus strains of *P. syringae*:

1. Digest *P. syringae*'s DNA with restriction enzymes.
2. Insert the individual DNA pieces into a plasmid. Pieces will insert randomly, allowing for different variations of recombinant DNA to be produced.
3. Transform the bacterium *Escherichia coli* (*E.coli*) with the recombinant plasmid. The plasmid will be taken in by the bacteria, rendering it part of the organism's DNA.
4. Identify the ice-gene from the numerous newly developed *E. coli* recombinants. Recombinant *E. coli* with the ice-gene will possess the ice-nucleating phenotype, these will be "ice-plus."
5. With the ice nucleating recombinant identified, amplify the ice gene with techniques such as polymerase chain reactions (PCR).
6. Create mutant clones of the ice gene through the introduction of mutagenic agents such as UV radiation to inactivate the ice gene, creating the "ice-minus" gene.
7. Repeat previous steps (insert gene into plasmid, transform *E. coli*, identify recombinants) with the newly

created mutant clones to identify the bacteria with the ice-minus gene. They will possess the desired ice-minus phenotype.

8. Insert the ice-minus gene into normal, ice-plus *P. syringae* bacterium.
9. Allow recombination to take place, rendering both ice-minus and ice-plus strains of *P. syringae*.

Economic Importance

The success of the agricultural world is heavily dependent on the weather. Cold weather conditions are directly responsible for the appearance of frost on plants and most importantly, crops. In the United States alone, it has been estimated that frost accounts for approximately $1 billion in crop damage each year. As *P. syringae* commonly inhabits plant surfaces, its ice nucleating nature incites frost development, freezing the buds of the plant and destroying the occurring crop. The introduction of an ice-minus strain of *P. syringae* to the surface of plants would incur competition between the strains. Should the ice-minus strain win out, the ice nucleate provided by *P. syringae* would no longer be present, lowering the level of frost development on plant surfaces at normal water freezing temperature (0°C). Even if the ice-minus strain does not win out, the amount of ice nucleate present from ice-plus *P. syringae* would be reduced due to competition. Decreased levels of frost generation at normal water freezing temperature would translate into a lowered quantity of crops lost due to frost damage, rendering higher crop yields overall.

Historical Perspective

Dr. Hall Hoppe of the US Department of Agriculture was the first to notice a connection between bacteria and frost damage. In 1961, Dr. Hoppe studied a corn fungus by grinding up infected leaves each season, then applying the powder to test corn for the following season to track the disease. A surprise frost occurred that year, leaving peculiar results. Only plants infected with the diseased powder incurred frost damage, leaving healthy plants unfrozen.

This phenomenon would baffle scientists until graduate student Stephen Lindow of the University of Wisconsin-Madison found a bacterium in the dried leaf powder in the early 1970s. Dr. Lindow, now a plant pathologist at the University of California-Berkeley, found that when this particular bacterium was introduced to plants where it is originally absent, the plants became very vulnerable to frost damage. He would go on to identify the bacterium as *P. syringae*, investigate *P. syringae*'s role in ice nucleation and in 1977, discover the mutant ice-minus strain. He was later successful at developing the ice-minus strain of *P. syringae* through recombinant DNA technology as well.

In 1983, Advanced Genetic Sciences (AGS) obtained US government authorisation to perform field tests with the ice-minus strain of *P. syringae*, but environmental groups and protestors delayed the field tests for four years with legal challenges. In 1987, the ice-minus strain of *P. syringae* became the first genetically modified organism (GMO) to be released into the environment. A strawberry field in California was spayed with the ice-minus strain of *P. syringae* just before a frost in 1987. The results were promising, showing lowered frost damage to the treated plants, but the data was in suspect as environment activists destroyed some of the plants. Dr. Lindow also conducted an experiment on a crop of potato seedlings sprayed with ice-minus *P. syringae*. He was successful in protecting the potato crop from frost damage with a strain of ice-minus *P. syringae*.

At the time of Dr. Lindow's work on ice-minus *P. syringae*, genetic engineering was considered to be very controversial. The controversy primarily revolved around fears of introducing new organisms that may permanently disrupt the ecosystem. The fear was that the introduction of ice-minus bacteria to the environment would eliminate bacterial and plant varieties. This was true in the case of the gypsy moth's accidental introduction into the US Without a predator in the US, the gypsy moth is still causing overwhelming destruction to the hardwood forests of northeastern US.

Ice Nucleating Bacteria: How recombinant DNA technology can protect plants from frost damage Ice nucleators: Did you think that water freezes at 32°F. In reality, pure water can be "super-cooled" to temperatures as low as -40°F before freezing occurs. Ice crystals are formed in cold temperatures, in the presence of moisture, and with the help of a "seed" to get the process started. This seed is known as an ice nucleator. The addition of an ice nucleator slows down the motion of water molecules, removing enough energy from the water to allow the molecules to align in a crystalline lattice and enter the solid state. An ice nucleator is any foreign particle, such as a piece of dust, which incites the freezing process.

Frost Damage: Frost is responsible for approximately $1 billion in crop losses each year in the United States. Bacteria living on plant surfaces are partly responsible for frost damage by serving as biological ice nuclei. *Pseudomonas syringae,* a common bacterium found on plants, in soil, and in the air, produces an ice-nucleation protein in the outermembrane of its cells. With *Pseudomonas syringae* inhabiting a citrus blossom, water can freeze at higher temperatures and effectively kill the buds. Many of us have heard about frost damage to Florida's citrus crop on the news.

Historical Perspective: What was the first recombinant microbe released into the environment? *Pseudomonas syringae.* This all started when scientists envisioned creating a strain of the bacterium which does not produce the ice nucleation protein. The idea was that an 'ice minus' mutant could be sprayed on crops in hopes that it would outcompete the naturally occurring ice positive strains. This is the first use of a genetically modified organism to biologically control another organism.

The Controversy: The genetic modification of *P. syringae* did not involve the addition of any genetic material, but rather, the deletion of a gene. Regardless, at the time of this work (early 1980s), any genetic modification of an organism was controversial. A company, Advanced Genetic Sciences, applied to the government to get permission to conduct field trials with the ice minus bacterium. In 1983, the company gained permission from the National Institute of Health (NIH) and the Environmental Protection Agency (EPA) to conduct field experiments with the ice minus strain, but environmental groups legally challenged these experiments. It wasn't until 1987 that the first field trials could be completed. The result: spraying a potato crop with the ice minus bacterium offered protection from frost.

Production of 'Ice Minus' Bacteria: To delete the ice gene from *P. syringae*, researchers first had to find and clone the gene. This was accomplished by cutting *P. syringae*'s DNA into pieces using restriction enzymes. The pieces of DNA were then inserted into a plasmid of *E. coli*. Because many different recombinants were produced, the *E. coli* containing the ice gene had to be selected. This is relatively simple since the *E. coli* with the ice gene will have the phenotype of an ice nucleator. Once researchers found and cloned the ice gene, they were able to make recombinant, 'ice minus' mutants of *P. syringae* with deletion of the ice gene.

Bacterial Artificial Chromosome

A bacterial artificial chromosome (BAC) is a DNA construct, based on a fertility plasmid (or F-plasmid), used for transforming and cloning in bacteria, usually *E. coli*. F-plasmids play a crucial role because they contain partition genes that promote the even distribution of plasmids after bacterial cell division. The bacterial artificial chromosome's usual insert size is 150 kbp, with a range from 100 to 300 kbp. A similar cloning vector, called a PAC has also been produced from the bacterial P1-plasmid.

BACs are often used to sequence the genetic code of organisms in genome projects, for example the Human Genome

Project. A short piece of the organism's DNA is amplified as an insert in BACs, and then sequenced. Finally, the sequenced parts are rearranged in silico, resulting in the genomic sequence of the organism.

Contribution to Models of Disease

BACs are now being utilised to a greater extent in modelling genetic diseases, often alongside transgenic mice. BACs have been useful in this field as complex genes may have several regulatory sequences upstream of the encoding sequence, including various promotion sequences that will show the genes true expression level (in a human for instance). BACs have been used to some degree of success with mice when studying neurological diseases such as Alzheimer's and aneuploidy with Down's syndrome. There have also been instances when they have been used to study specific oncogenes associated with cancers. They are transferred over to these genetic disease models by electroporation/transformation, transfection with a suitable virus or microinjection. BACs can also be utilised to detect genes or large sequences of interest and then used to map them onto the human chromosome using BAC arrays. BACs are preferred for these kind of genetic studies because they tend to contain a much larger, more stable sequence than other types of cloning vectors.

8

Biotechnological Applications of Cynobacteria

Cyanobacteria (Greek) = blue + bacterium) also known as Cyanophyta is a phylum (or "division") of Bacteria that obtain their energy through photosynthesis. They are often referred to as blue-green algae, although they are in fact prokaryotes, not algae. The description is primarily used to reflect their appearance and ecological role rather than their evolutionary lineage. The name "cyanobacteria" comes from the colour of the bacteria, cyan; the bacteria do not use or produce cyanide whose chemical prefix is cyano.

Putative fossil traces of cyanobacteria have been found from around 3.8 billion years ago (bya). They are a significant component of the marine nitrogen cycle and an important primary producer in many areas of the ocean.

Their ability to perform oxygenic (plant-like) photosynthesis is thought to have converted the early reducing atmosphere into an oxidising one, which dramatically changed the life forms on Earth and provoked an explosion of biodiversity.

Various Forms

Cyanobacteria are found in almost every conceivable habitat, from oceans to fresh water to bare rock to soil. Most are found in fresh water, while others are marine, occur in damp soil, or even temporarily moistened rocks in deserts. A few are endosymbionts in lichens, plants, various protists, or sponges and provide energy for the host. Some live in the fur of sloths, providing a form of camouflage.

Cyanobacteria include unicellular and colonial species. Colonies may form filaments, sheets or even hollow balls. Some filamentous colonies show the ability to differentiate into several different cell types: vegetative cells, the normal, photosynthetic cells that are formed under favourable growing conditions; akinetes, the climate-resistant spores that may form when environmental conditions become harsh; and thick-walled heterocysts, which contain the enzyme nitrogenase, vital for nitrogen fixation.

Heterocysts may also form under the appropriate environmental conditions (anoxic) wherever nitrogen is necessary. Heterocyst-forming species are specialised for nitrogen fixation and are able to fix nitrogen gas, which cannot be used by plants, into ammonia (NH_3), nitrites (NO_2^-) or nitrates (NO_3^-), which can be absorbed by plants and converted to protein and nucleic acids. The rice paddies of Asia, which produce about 75 per cent of the world's rice, could not do so were it not for healthy populations of nitrogen-fixing cyanobacteria in the rice paddy fertilizer.

Many cyanobacteria also form motile filaments, called hormogonia, that travel away from the main biomass to bud and form new colonies elsewhere. The cells in a hormogonium are often thinner than in the vegetative state, and the cells on either end of the motile chain may be tapered. In order to break away from the parent colony, a hormogonium often must tear apart a weaker cell in a filament, called a necridium.

Each individual cell of a cyanobacterium typically has a

thick, gelatinous cell wall, which stains gram-negative. They lack flagella, but hormogonia and some unicellular species may move about by gliding along surfaces. In water columns some cyanobacteria float by forming gas vesicles, like in archaea.

Photosynthesis

Cyanobacteria have an elaborate and highly organised system of internal membranes which function in photosynthesis. Photosynthesis in cyanobacteria generally uses water as an electron donor and produces oxygen as a by-product, though some may also use hydrogen sulphide as occurs among other photosynthetic bacteria. Carbon dioxide is reduced to form carbohydrates via the Calvin cycle. In most forms the photosynthetic machinery is embedded into folds of the cell membrane, called thylakoids.

The large amounts of oxygen in the atmosphere are considered to have been first created by the activities of ancient cyanobacteria. Due to their ability to fix nitrogen in aerobic conditions they are often found as symbionts with a number of other groups of organisms such as fungi (lichens), corals, pteridophytes (Azolla), angiosperms (Gunnera), etc.

Cyanobacteria are the only group of organisms that are able to reduce nitrogen and carbon in aerobic conditions, a fact that may be responsible for their evolutionary and ecological success. The water-oxidising photosynthesis is accomplished by coupling the activity of photosystem (PS) II and I (Z-scheme). In anaerobic conditions, they are also able to use only PS I — cyclic photophosphorylation — with electron donors other than water (hydrogen sulphide, thiosulphate, or even molecular hydrogen) just like purple photosynthetic bacteria. Furthermore, they share an archaebacterial property, which is the ability to reduce elemental sulphur by anaerobic respiration in the dark. Perhaps the most intriguing thing about these organisms is that their photosynthetic electron transport shares the same compartment as the components of respiratory electron transport. Actually, their plasma membrane contains only components of the respiratory chain, while the thylakoid membrane hosts both respiratory and photosynthetic electron transport.

Attached to thylakoid membrane, phycobilisomes act as light harvesting antennae for the photosystems. The phycobilisome components (phycobiliproteins) are responsible for the blue-green pigmentation of most cyanobacteria. The variations to this theme is mainly due to carotenoids and phycoerythrins which give the cells the red-brownish colouration. In some cyanobacteria, the colour of light influences the composition of phycobilisomes. In green light, the cells accumulate more phycoerythrin, whereas in red light they produce more phycocyanin.

Thus the bacteria appear green in red light and red in green light. This process is known as complementary chromatic adaptation and is a way for the cells to maximise the use of available light for photosynthesis. A few genera, however, lack phycobilisomes and have chlorophyll *b* instead (*Prochloron, Prochlorococcus, Prochlorothrix*). These were originally grouped together as the prochlorophytes or chloroxybacteria, but appear to have developed in several different lines of cyanobacteria. For this reason they are now considered as part of cyanobacterial group.

Relationship to Chloroplasts

Chloroplasts found in eukaryotes (algae and higher plants) likely evolved from an endosymbiotic relation with cyanobacteria. This endosymbiotic theory is supported by various structural and genetic similarities. Primary chloroplasts are found among the green plants, where they contain chlorophyll *b*, and among the red algae and glaucophytes, where they contain phycobilins. It now appears that these chloroplasts probably had a single origin, in an ancestor of the clade called Primoplantae. Other algae likely took their chloroplasts from these forms by secondary endosymbiosis or ingestion.

It was once thought that the mitochondria in eukaryotes also developed from an endosymbiotic relationship with cyanobacteria; however, we now suspect that this evolutionary event occurred when aerobic Eubacteria were engulfed by anaerobic host cells. Mitochondria are believed to have originated not from cyanobacteria but from an ancestor of Rickettsia.

Classification

The cyanobacteria were traditionally classified by morphology into five sections, referred to by the numerals I-V. The first three—Chroococcales, Pleurocapsales, and Oscillatoriales—are not supported by phylogenetic studies. However, the latter two—Nostocales and Stigonematales—are monophyletic, and make up the heterocystous cyanobacteria. The members of Chroococales are unicellular and usually aggregated in colonies. The classic taxonomic criterion has been the cell morphology and the plane of cell division. In Pleurocapsales, the cells have the ability to form internal spores (baeocytes). In Oscillatorialles, the cells are uniseriately arranged and do not form specialised cells (akinets and heterocysts). In Nostocalles and Stigonematalles the cells have the ability to develop heterocysts in certain conditions. Stigonematales, unlike Nostocalles include species with truly branched trichome. Most taxa included in the phylum or division Cyanobacteria have not yet been validly published under the Bacteriological Code. Except:

- The classes Chroobacteria, Hormogoneae and Gloeobacteria.
- The orders Chroococcales, Gloeobacterales, Nostocales, Oscillatoriales, Pleurocapsales and Stigonematales.
- The families Prochloraceae and Prochlorotrichaceae.
- The genera Halospirulina, Planktothricoides, Prochlorococcus, Prochloron, Prochlorothrix.

Biotechnology and Applications

Certain cyanobacteria produce cyanotoxins like Anatoxin-a, Anatoxin-as, Aplysiatoxin, Cylindrospermopsin, Domoic acid, Microcystin LR, Nodularin R (from *Nodularia*), or Saxitoxin. Sometimes a mass-reproduction of cyanobacteria results in algal blooms.

The unicellular cyanobacterium *Synechocystis* sp. PCC 6803 was the first photosynthetic organism whose genome was completely sequenced (in 1996, by the Kazusa Research Institute, Japan). It continues to be an important model organism.

At least one secondary metabolite, cyanovirin, has shown to possess anti-HIV activity.

Some cyanobacteria are sold as food, notably *Aphanizomenon flos-aquae* (E3live) and *Arthrospira platensis* (Spirulina). It has been suggested that they could be a much more substantial part of human food supplies, as a kind of superfood.

Along with algae, some hydrogen producing cyanobacteria are being considered as an alternative energy source, notably at Oregon State University, in research supported by the US Department of Energy, Princeton University, Colorado School of Mines as well as at Uppsala University, Sweden

Anatoxin-α

Anatoxin-a is a secondary, bicyclic amine alkaloid with acute neurotoxicity. The toxin is produced by at least four different genera of cyanobacteria and has been reported in North America, Europe, Africa, Asia, and Australia. Symptoms of anatoxin exposure are loss of coordination, muscular fasciculations, convulsions and death by respiratory paralysis. It's mode of action is through the nicotinic acetylcholine receptor where it acts as an analogue of acetylcholine. The molecule is not degraded by cholinesterase thus causes permanent stimulation of muscle cells leading to paralysis. Information on carcinogenicity is not available.

Cylindrospermopsin

Cylindrospermopsin is a toxin of the cyanobacteria Cylindrospermopsin raciborskii or Microcystis aeruginosa. It is a polycyclic uracil derivative produced by a variety of freshwater cyanobacteria. Several slightly different variants (deoxycylindro-spermopsin and 7-epicylindrospermopsin) are also known to be produced by cyanobacteria. The molecule exhibits toxicity to liver and kidney tissue and is thought to inhibit protein sysnthesis and to covalently (itself or a metabolite) binding to DNA and/or RNA. The molecule has been observed to be produced by *Cylindrospermopsis raciborskii, Umezakia natans, Aphanizomenon ovalisporum, Anabaena bergii,* and *Raphidiopsis curvata*. The carcinogenicity of cylindrospermopsin has not been determined but it appears to have no tumor initiating activity in mice.

Microcystin

Microcystins are cyclic non-ribosomal peptides produced by cyanobacteria. They are cyanotoxins and can be very toxic for plants and animals including humans. Their hepatotoxicity may cause serious damage to the liver.

Microcystins consist of several uncommon non-proteinogenic amino acids such as dehydroalanine derivatives and the special a-amino acid ADDA (*(all-S,all-E)*-3-Amino-9-methoxy-2,6,8-trimethyl-10-phenyldeca-4,6-diene acid).

There appears to be inadequate information to assess carcinogenic potential of microcystins by applying EPA Guidelines for Carcinogen Risk Assessment. A few studies suggest that there may be a relationship between liver and colorectral cancers and the occurrence of cyanobacteria in drinking water in China (Yu *et al.*, 1989; Zhou *et al.*, 2002). Evidence is, however, limited due to limited ability to accurately assess and measure exposure.

Saxitoxin

Saxitoxin (STX) is a neurotoxin found in marine dinoflagellates (algae). It is a selective sodium channel blocker. The United States military isolated saxitoxin and assigned it the chemical weapon designation TZ. It is unique among toxins in that it acts in a matter of minutes. The LCt_{50} of TZ is 5 mg·min/m³.

The medical importance of saxitoxin is in relation to red tide in shellfish and causes the paralytic shellfish poisoning (PSP) food poisoning. The blocking of the sodium channel produces a flaccid paralysis that leaves its victim calm and conscious through the progression of symptoms. Death is from respiratory failure.

It is listed in schedule 1 of the Chemical Weapons Convention. Though its early isolation and characterisation were related to military efforts, saxitoxin has been more important to cellular research in describing the function of the sodium channel.

Nitrogen Assimilation and Nitrogen Control in Cyanobacteria

Nitrogen sources commonly used by cyanobacteria include ammonium, nitrate, nitrite, urea and atmospheric N_2, and some cyanobacteria can also assimilate arginine or glutamine. ABC (ATP-binding cassette)-type permeases are involved in the uptake of nitrate/nitrite, urea and most amino acids, whereas secondary transporters take up ammonium and, in some strains, nitrate/nitrite. In cyanobacteria, nitrate and nitrite reductases are ferredoxin-dependent enzymes, arginine is catabolised by a combination of the urea cycle and arginase pathway, and urea is degraded by a Ni^{2+} -dependent urease. These pathways provide ammonium that is incorporated into carbon skeletons through the glutamine synthetase–glutamate synthase cycle, in which 2-oxoglutarate is the nal nitrogen acceptor. The expression of many nitrogen assimilation genes is subjected to regulation being activated by the nitrogen-control transcription factor NtcA, which is autoregulatory and whose activity appears to be influenced by 2-oxoglutarate and the signal transduction protein PII. In some lamentous cyanobacteria, N_2 fixation takes place in specialised cells called heterocysts that differentiate from vegetative cells in a process strictly controlled by NtcA.

The oxygenic photosynthetic bacteria, the cyanobacteria,

are autotrophs that fix CO_2 through the reductive pentose phosphate cycle. They make an important contribution to the Earth's nitrogen cycle by incorporating nitrogen into the biosphere through assimilatory processes. Organisms of this group frequently have the capacity to assimilate as sources of nitrogen a number of different simple N-containing compounds including ammonium, nitrate, nitrite and urea. Many cyanobacteria can also assimilate atmospheric nitrogen (a process known as N_2 fixation), and still some strains have the capability to assimilate some amino acids, particularly arginine and glutamine (reviewed in). The assimilation of most of these compounds provides intracellular ammonium, which itself is a preferred nitrogen source.

Thus, in the presence of ammonium, the genes encoding permeases and enzymes for the assimilation of nitrogen sources alternative to ammonium are repressed, a process known as 'nitrogen control'. Within the context of unravelling the mechanisms of nutrient incorporation in the biosphere, the aim of our research is to draw a complete picture of the mechanisms of nitrogen assimilation and its regulation in cyanobacteria.

Mechanisms of Assimilation of Combined Nitrogen

The incorporation into the cell of nitrogen-containing compounds, which are frequently found at low concentrations (e.g. below 1 μM) in the environment, takes place through permeases that are located in the cytoplasmic membrane. Multicomponent ABC (ATP-binding cassette)-type uptake transporters have been shown to be involved in the uptake of nitrate and nitrite or urea in a number of cyanobacteria. ABC-type permeases are also required for the transport of arginine and glutamine. These permeases use ATP to drive an active, concentrative transport of their substrates. On the other hand, a secondary transporter of the major facilitator superfamily has been identified as the nitrate–nitrite transporter in some marine cyanobacteria.

The transport of ammonium is also mediated by secondary

permeases, in this case of the Amt family. The Amt permeases can be probed with methylammonium, which has been shown to be concentrated in cells of the unicellular cyanobacterium *Synechocystis* sp. Intracellular nitrate is sequentially reduced to nitrite and ammonium by nitrate reductase and nitrite reductase, which are the products of the *narB* and *nir* genes respectively. Cyanobacterial nitrite reductase is homologous with ferredoxin-dependent higher-plant nitrite reductase and contains a [4Fe-4S] cluster and sirohaem as prosthetic groups. Electrons from reduced ferredoxin are transferred to the iron–sulphur cluster and then to sirohaem, where nitrite is reduced to ammonium.

Cyanobacterial nitrate reductase is homologous with Mo-containing bacterial oxidoreductases but is unique in that it uses ferredoxin as an electron donor, forming tight 1:1 complexes. The Mo cofactor is of the Mo-*bis*-molybdopterin guanine dinucleotide type, and the enzyme also contains a [4Fe-4S] cluster. In this enzyme system, electrons flow from reduced ferredoxin to the iron–sulphur cluster and then to the Mo cofactor, where nitrate is reduced to nitrite. The *narB* and *nir* genes are clustered together with the nitrate/nitrite permease-encoding genes in numerous cyanobacteria forming an operonwith the structure *nir*-permease genes-*narB*. The high conservation of this gene arrangement, in which the expression level is higher for the upstream than for the downstream genes in the operon, suggests that it ensures the production of a balanced amount of the different proteins of the pathway.

Regarding organic sources of nitrogen used by cyanobacteria, urea is degraded to ammonium and CO_2 by a standard bacterial Ni^{2+} -dependent urease, whereas arginine is catabolised by an unusual pathway that combines the urea cycle and the arginase pathway rendering ammonium and glutamate as final products. Whatever the nitrogen source used for growth, intracellular ammonium is incorporated into carbon skeletons through the glutamine synthetase–glutamate synthase pathway (reviewed in). In cyanobacteria, which lack

2-oxoglutarate dehydrogenase, the main metabolic role of 2-oxoglutarate is the incorporation of nitrogen. This metabolic arrangement makes 2-oxoglutarate an indicator of the C to N ratio of the cells.

However, not all these pathways are present at the same time in a cyanobacterial cell, their expression being strictly regulated by the nitrogen source and also by the availability of carbon.

Nitrogen Control

When the cyanobacterial cells are incubated in the presence of a limiting concentration of ammonium, but with an adequate supply of carbon, they sense a high C to N ratio that determines expression of genes encoding permeases and enzymes required for an efficient assimilation of ammonium or for the assimilation of alternative nitrogen sources.

This activation of gene expression requires NtcA, a transcriptional regulator that belongs to the family of bacterial transcription factors whose best-known representative is CAP (catabolite gene activator protein) of *Escherichia coli*. Modelling of NtcA based on the CAP structure and preliminary x-ray structural data confirms that NtcA bears a structure similar to that of CAP, bearing a DNA-binding helix–turn–helix motif in its C-terminal domain. NtcA binds to specific sites in the promoters of the regulated genes activating their transcription. The DNA-binding site of NtcA shows the sequence signature GTAN8TAC that is usually found 22 nt upstream from a promoter-10 box. This promoter structure for NtcA-activated genes, which conforms to the structure of bacterial class II promoters, has been found in most cyanobacterial nitrogen-regulated genes investigated to date.

The NtcA gene is autoregulatory and the NtcA protein appears to be subjected to regulation being activated when the C to N ratio of the cell is high. Although the precise mechanism of modulation of NtcA is still not known, two elements that influence NtcA activity have recently been identified:

2-oxoglutarate and the PII protein. Binding of NtcA to some N-regulated promoters is stimulated by 2-oxoglutarate, suggesting that this metabolite can represent an allosteric effector for NtcA. The cyanobacterial signal transduction PII protein (*glnB* gene product) binds ATP and 2-oxoglutarate and is phosphorylated when the cells experience a high C to N ratio, it having been suggested that it operates as a 2-oxoglutarate sensor.

The PII protein is required for the activation of NtcAdependent genes specifically under conditions of nitrogen stress, and it appears to promote gene expression more efficiently in its phosphorylated form. Therefore 2-oxoglutarate may affect NtcA activity directly and indirectly through PII. *In vivo* positive effects of 2-oxoglutarate on the expression of nitrogen-regulated genes have been observed in an engineered *Synechococcus* strain bearing the *E. coli* 2-oxoglutarate permease.

Some NtcA-dependent promoters appear to require additional regulators or transcription factors for proper operation. This is the case of the *Anabaena nir* operon promoter that requires binding of NtcB (a LysR-family transcription factor) in addition to binding of NtcA. Use of this promoter is also dependent on CnaT, a putative glycosyl transferase that influences transcription through a mechanism that is yet to be worked out. In addition to the regulation of gene expression, some elements of the nitrogen assimilation system are subjected to post-translational regulation in cyanobacteria. The nitrate/ nitrite permease is a target of regulation, being inhibitedwhen the cells sense a high N to C ratio. The non-phosphorylated form of the PII protein appears to effect specific inhibition of the permease when the cells are incubated under conditions that determine low 2-oxoglutarate levels.

Additionally, in the unicellular cyanobacterium *Synechocystis* sp. strain PCC 6803, glutamine synthetase is subjected to cell wall, which carries extra glycolipid and polysaccharide layers and shows a low permeability to gases.

The heterocysts also exhibit a modified metabolism oriented towards keeping a microaerobic environment for nitrogenase expression and towards providing ATP and electrons for nitrogenase function.

NtcA is strictly required for heterocyst development, apparently linking differentiation to nitrogen deficiency, and is needed for the induction of the key heterocyst development regulator HetR. Moreover, we have shown thatNtcA also participates in the activation of expression of geneswhose products are required during the differentiation process or are active in the mature heterocyst. Early developmental genes such as the *hetC* regulatory gene and genes that are expressed later during the developmental process such as *devBCA*, encoding a glycolipid exporter, or the two *cox* operons, encoding cytochrome oxidases that are essential for N_2 fixation, are transcribed from NtcAdependent promoters. In the mature heterocyst, the *glnA* gene encoding glutamine synthetase, *petH* encoding ferredoxin:$NADP^+$ reductase and the *nifHDK* operon encoding the nitrogenase complex are also expressed from NtcA-activated promoters. We have recently shown that genes for the metabolism of cyanophycin, a nitrogen reserve made of aspartate and arginine that accumulates conspicuously in the heterocysts, are expressed from multiple promoters also including NtcA-dependent promoters.

Some of the heterocyst-related NtcA-dependent promoters do not present the structure of the standard, class II NtcA-type promoter, but rather show the structure of bacterial class I promoters with the NtcA-binding site being located approx. 90 pre cent upstream from the transcription start point. In summary, NtcA effects nitrogen control not only for triggering the developmental process but also for its progression and in the N_2-fixing heterocyst, and does so through the use of different types of NtcA-dependent promoters.

The N_2-fixing heterocyst provides the vegetative cells of the filament with fixed nitrogen, probably in the form of amino acids. Glutamine is probably a transferred amino acid, but whether cyanophycin amino acids have a role in the nitrogen transfer from the heterocysts to the vegetative cells is currently not known. The mechanism of transfer is also not known, but mutants of some amino acid permeases of *Anabaena* sp. strain PCC 7120 have been shown to be impaired in diazotrophic growth, suggesting a role for such amino acid permeases in intercellular nitrogen transfer in the N_2-fixing filament.

Cyanobacteria and Carotenoids: Return to Harmony with Nature

The fact that carotenes and even carotenoids, and in some cases secondary carotenoids, are beneficial to human health is unmistakeable and scientifically proven. It is necessary to emphasise though that it doesn't directly cure but that its strength lies in its preventative effects. This means we need to balance our intake of necessary substances in order to maintain our health at a certain level or to overcome any ailments. This is where carotenoids are invaluable. Where it concerns defence mechanisms against free radicals, which can be very destructive at cell level, it strengthens the immune system and in some cases has a whole range of effects connected not only with internal workings but also a whole range of protective effects against harmful elements of the Sun's radiation.

A frequently debated question is the source of these health enhancing substances. It is quite common to use carotenoids and similar substances in their natural form, isolated from plant material, in some cases from cyanobacteria, cultivated for this purpose. The yield in this case is small but the material obtained, after a little alteration, is more biologically balanced and has shown to have a synergistic effect, i.e. this balancing effect is carried through in natural materials containing molecules whose function we do not yet know about. This is a big and often neglected advantage of natural unrefined products.

Cheaper products are prepared these days with the aid of molecular biological methods. These are now unprecedentedly widespread and thanks to these methods sigificant advances have been made in the production of medicines. Where this concerns food supplements, in our case carotenoids, understandably, molecular biology has expanded considerably into this field too. The principle employed here is to introduce a sequence of genes which create a product (in our case carotenoids) in greater amounts, but only as a single substance (often further purified and isolated) with a quantative imbalance with the other necessary molecules. All things should be in harmony and so food supplements (and not only supplements) created by means of genetic manipulation are unsuited to this requirement for balance. Convincing evidence of this is currently being put forward by the scientific branch known as metabolics.

The cheapest food supplements are constructed completely artificially, using chemical methods. Any thought of balance in natural carotenoids in this case is not even a consideration. At this point it is essential to mention other enriching molecules which are welcome supplements in illness prevention. At the current time selenium is often mentioned for its effect in improving memory and preventing certain illnesses connected with brain capacity. Here too we must again ask the question of what is the appropriate form which should be contained in the resulting product. As a pure mineral, as a sulphate or chloride salt it has no significance for humans. If however the selenium is bound organically it is accepted as a precurser to certain proteins and even enzymes, where for example, selenium is substituted into the amino acid sulphur methionine then the organism takes up this substance in a form easy to absorb and transport via natural biochemical pathways. Just such bound selenium can be obtained from cultivated cyanobacteria. This product has a synergistic effect with the accompanying molecule.

Finally it is possible to add just one more thing. Natural substances, without any kind of genetic manipulation, or mere chemical additives, have one immeasurable advantage. This is the synergistic effect of the additional molecule which helps uptake and exploitation of the biologically active substances by cells and, of course, the organism.

Cyanobacteria in the Environment

Nature and Diversity

Plants and animals possess consistent features by which they can be identified reliably and sorted into recognisably distinct groups. Biologists observe and compare what the organisms look like, how they grow and what they do. The results make it possible to construct systematic groupings based on multiple correlations of common characters and that reflect the greatest overall similarity. The basis for such groupings is the fact that all organisms are related to one another by way of evolutionary descent. Their biology and phylogenetic relationships makes the establishment of systematic groupings possible (Minkoff, 1983).

However, microbial systematics has long remained an enigma. Conceptual advances in microbiology during the twentieth century included the realisation that a discontinuity exists between those cellular organisms that are prokaryotic (i.e. whose cells have no nucleus) and those that are eukaryotic (i.e. more complexly structured cells with a nucleus) within the organisation of their cells. The microalgae investigated by phycologists under the International Code of Botanical Nomenclature (ICBN) (Greuter *et al.*, 1994) included organisms of both eukaryotic and prokaryotic cell types. The bluegreen algae (Geitler, 1932) constituted the largest group of the latter category. The prokaryotic nature of these organisms and their fairly close relationship with eubacteria made work under provisions of the International Code of Nomenclature of Bacteria (ICNB) (Sneath, 1992) more appropriate (Rippka *et al.*, 1979; Waterbury, 1992).

The prevailing systematic view is that comparative studies of the genetic constitution of the cyanobacteria will now contribute significantly to the revision of their taxonomy. Relevant classification should reflect as closely as possible the phylogenetic relationships as, for example, encoded in 16S or 23S rRNA sequence data (Woese, 1987). The integration of phenotypic, genotypic and phylogenetic information render possible a consensus type of taxonomy known as polyphasic taxonomy (Vandamme *et al.*, 1996).

The names "cyanobacteria" and "blue-green algae" (Cyanophyceae) are valid and compatible systematic terms. This group of microorganisms comprises unicellular to multicellular prokaryotes that possess chlorophyll *a* and perform oxygenic photosynthesis associated with photosystems I and II (Castenholz and Waterbury, 1989).

Occurrence in Nature

The majority of cyanobacteria are aerobic photoautotrophs. Their life processes require only water, carbon dioxide, inorganic substances and light. Photosynthesis is their principal mode of energy metabolism. In the natural environment, however, it is known that some species are able to survive long periods in complete darkness. Furthermore, certain cyanobacteria show a distinct ability for heterotrophic nutrition (Fay, 1965). Cyanobacteria are often the first plants to colonise bare areas of rock and soil.

Adaptations, such as ultraviolet absorbing sheath pigments, increase their fitness in the relatively exposed land environment. Many species are capable of living in the soil and other terrestrial habitats, where they are important in the functional processes of ecosystems and the cycling of nutrient elements (Whitton, 1992).

The prominent habitats of cyanobacteria are limnic and marine environments. They flourish in water that is salty, brackish or fresh, in cold and hot springs, and in environments where no other microalgae can exist. Most marine forms (Humm and Wicks, 1980) grow along the shore as benthic vegetation in the zone between the high and low tide marks. The cyanobacteria comprise a large component of marine plankton with global distribution (Wille, 1904; Gallon *et al.*,

1996). A number of freshwater species are also able to withstand relatively high concentrations of sodium chloride. It appears that many cyanobacteria isolated from coastal environments tolerate saline environments (i.e. are halotolerant) rather than require salinity (i.e. are halophilic). As frequent colonisers of euryhaline (very saline) environments, cyanobacteria are found in salt works and salt marshes, and are capable of growth at combined salt concentrations as high as 3-4 molar mass (Reed *et al.,* 1984). Freshwater localities with diverse trophic states are the prominent habitats for cyanobacteria. Numerous species characteristically inhabit, and can occasionally dominate, both near-surface epilimnic and deep, euphotic, hypolimnic waters of lakes (Whitton, 1973). Others colonise surfaces by attaching to rocks or sediments, sometimes forming mats that may tear loose and float to the surface. Cyanobacteria have an impressive ability to colonise infertile substrates such as volcanic ash, desert sand and rocks (Jaag, 1945; Dor and Danin, 1996). They are extraordinary excavators, boring hollows into limestone and special types of sandstone (Weber *et al.,* 1996). Another remarkable feature is their ability to survive extremely high and low temperatures. Cyanobacteria are inhabitants of hot springs (Castenholz, 1973), mountain streams (Kann, 1988), Arctic and Antarctic lakes (Skulberg, 1996a) and snow and ice (Kol, 1968; Laamanen, 1996). The cyanobacteria also include species that run through the entire range of water types, from polysaprobic zones to katharobic waters (Van Landingham, 1982).

Cyanobacteria also form symbiotic associations with animals and plants. Symbiotic relations exist with, for example, fungi, bryophytes, pteridophytes, gymnosperms and angiosperms (Rai, 1990). The hypothesis for the endosymbiotic origin of chloroplasts and mitochondria should be mentioned in this context. The evolutionary formation of a photosynthetic eukaryote can be explained by a cyanobacteria being engulfed and codeveloped by a phagotrophic host (Douglas, 1994).

Fossils of what were almost certainly prokaryotes are

present in the 3,450 million year old Warrawoona sedimentary rock of northwestern Australia. Cyanobacteria were among the pioneer organisms of the early earth (Brock 1973; Schopf, 1996). These photosynthetic microorganisms were, at that time, probably the chief primary producers of organic matter, and the first organisms to release elemental oxygen into the primitive atmosphere. Sequencing of deoxyribonucleic acid (DNA) has given evidence that the earliest organisms were thermophilic and thus able to survive in oceans that were heated by volcanoes, hot springs and bolide impacts (Holland, 1997).

Organisation, Function and Behaviour

The structure and organisation of cyanobacteria are studied using light and electron microscopes. The basic morphology comprises unicellular, colonial and multicellular filamentous forms.

Unicellular forms, for example in the order Chroococcales, have spherical, ovoid or cylindrical cells. They occur singly when the daughter cells separate after reproduction by binary fission. The cells may aggregate in irregular colonies, being held together by the slimy matrix secreted during the growth of the colony. By means of a more or less regular series of cell division, combined with sheath secretions, more ordered colonies may be produced.

A particular mode of reproduction, which may supplement binary fission, distinguishes cyanobacteria in the order Chamaesiphonales and Pleurocapsales. In the Chamaesiphonales exospores are budded off from the upper ends of cells. In the second order, the principal mode of replication is by a series of successive binary fissions converting a single mother cell into many minute daughter cells (baeocytes or endospores).

Filamentous morphology is the result of repeated cell divisions occurring in a single plane at right angles to the main axis of the filament. The multicellular structure consisting of a chain of cells is called a trichome. The trichome may be straight or coiled.

Cell size and shape show great variability among the filamentous cyanobacteria. Species in the order Oscillatoriales, with unseriated and unbranched trichomes, are composed of essentially identical cells. The other orders with a filamentous organisation (orders Nostocales and Stigonematales) are characterised with trichomes having a heterogeneous cellular composition. Vegetative cells may be differentiated into heterocysts (having a thick wall and hyaline protoplast, capable of nitrogen fixation) and akinetes (large thick-walled cells, containing reserve materials, enabling survival under unfavourable conditions). In the order Stigonematales, the filaments are often multiseriated, with genuine branching. Both heterocysts and akinetes are present.

The only means of reproduction in cyanobacteria is asexual. Filamentous forms reproduce by trichome fragmentation, or by formation of special hormogonia. Hormogonia are distinct reproductive segments of the trichomes. They exhibit active gliding motion upon their liberation and gradually develop into new trichomes. In contrast to eukaryotic microalgae, cyanobacteria do not possess membrane-bound subcellular organelles; they have no discrete membrane-bound nucleus; they possess a wall structure based upon a peptidoglycan layer; and they contain 70 S rather than 80 S ribosomes (Fay and Van Baalen, 1987; Bryant, 1994).

The photosynthetic pigments of cyanobacteria are located in thylakoids that lie free in the cytoplasm near the cell periphery. Cell colours vary from blue-green to violet-red. The green of chlorophyll *a* is usually masked by carotenoids (e.g. beta-carotene) and accessory pigments such as phycocyanin, allophycocyanin and phycoerythrin (phycobiliproteins). The pigments are embodied in phycobilisomes, which are found in rows on the outer surface of the thylakoids (Douglas, 1994). All cyanobacteria contain chlorophyll *a* and phycocyanine.

The basic features of photosynthesis in cyanobacteria have been well described (Ormerod, 1992). Cyanobacteria are oxygenic phototrophs possessing two kinds of reaction centres,

PS I and PS II, in their photosynthetic apparatus. With the accessory pigments, they are able to use effectively that region of the light spectrum between the absorption peaks of chlorophyll *a* and the carotenoids. The ability for continuous photosynthetic growth in the presence of oxygen, together with having water as their electron donor for CO_2 reduction, enables cyanobacteria to colonise a wide range of ecological niches (Whitton, 1992). Phycobiliprotein synthesis is particularly susceptible to environmental influences, especially light quality. Chromatic adaptation is largely attributable to a change in the ratio between phycocyanin and phycoerythrin in the phycobilisomes.

Thus, cyanobacteria are able to produce the accessory pigment needed to absorb light most efficiently in the habitat in which they are present. Cyanobacteria have a remarkable ability to store essential nutrients and metabolites within their cytoplasm. Prominent cytoplasmic inclusions for this purpose can be seen with the electron microscope (e.g. glycogen granules, lipid globules, cyanophycin granules, polyphosphate bodies, carboxysomes) (Fay and Van Baalen, 1987).

Reserve products are accumulated under conditions of an excess supply of particular nutrients. For example, when the synthesis of nitrogenous cell constituents is halted because of an absence of a usable nitrogen source, the primary products of photosynthesis are channelled towards the synthesis and accumulation of glycogen and lipids.

Dinitrogen fixation is a fundamental metabolic process of cyanobacteria, giving them the simplest nutritional requirements of all living organisms. By using the enzyme nitrogenase, they convert N_2 directly into ammonium (NH_4) (a form through which nitrogen enters the food chain) and by using solar energy to drive their metabolic and biosynthetic machinery, only N_2, CO_2, water and mineral elements are needed for growth in the light. Nitrogen-fixing cyanobacteria are widespread among the filamentous, heterocyst forming genera (e.g. *Anabaena, Nostoc)* (Stewart, 1973). However, there are also several well documented examples of dinitrogen fixation among cyanobacteria not forming heterocysts (e.g. *Trichodesmium)* (Carpenter *et al.*, 1992). Under predominantly nitrogen limited conditions, but when other nutrients are available, nitrogen fixing cyanobacteria may be favoured and gain growth and reproductive success. Mass developments (often referred to as "blooms") of such species in limnic and marine environments (e.g. the Baltic Sea) are common phenomena worldwide.

Many species of cyanobacteria possess gas vesicles. These are cytoplasmic inclusions that enable buoyancy regulation and are gas-filled, cylindrical structures. Their function is to give planktonic species an ecologically important mechanism enabling them to adjust their vertical position in the water column (Walsby, 1987). To optimise their position, and thus to find a suitable niche for survival and growth, cyanobacteria use different environmental stimuli (e.g. photic, gravitational, chemical, thermal) as clues.

Gas vesicles become more abundant when light is reduced and the growth rate slows down. Increases in the turgor pressure of cells, as a result of the accumulation of photosynthate, cause a decrease in existing gas vesicles and therefore a reduction in buoyancy. Cyanobacteria can, by such buoyancy regulation, poise themselves within vertical gradients of physical and chemical factors. Other ecologically significant mechanisms of movement shown by some cyanobacteria are photomovement by slime secretion or surface undulations of cells (Hader, 1987; Paerl, 1988).

The presence of very small cells of cyanobacteria (in the size range 0.2-2 μm) has been recognised as a potentially significant source of primary production in various freshwater and marine environments. These cyanobacteria constitute a component of the picoplankton in pelagic ecosystems. Cells can be recognised and estimates of their abundance made by using epifluorescence microscopy (e.g. observing the orange fluorescence due to phycoerythrin). The unicellular genus

Synechococcus is one of the most studied, and geographically most widely distributed, cyanobacteria in the picoplankton. Toxigenic strains of *Synechococcus* have been reported (Skulberg *et al.*, 1993).

Biological Diversity

Although cyanobacteria probably evolved as a group of organisms about 2,000 million years before the advent of eukaryotes, they comprise fewer taxa than eukaryotic microalgae (Bisby, 1995). The concept of species in the cyanobacteria has, however, no distinct boundaries. The situation is similar for most organisms, except for those that are sexually reproductive. Depending on the classification system used, the number of species recognised varies greatly. Based on the International Code of Botanical Nomenclature the class Cyanophyceae, for example, contains about 150 genera and 2,000 species (Hoek *et al.*, 1995).

Chemotaxonomic studies include the use of markers, such as lipid composition, polyamines, carotenoids and special biochemical features. The resulting data support the more traditional examinations of phenotypic and ecological characteristics. Physiological parameters are conveniently studied using laboratory cultures (Packer and Glazer, 1988).

The diversity of cyanobacteria can be seen in the multitude of structural and functional aspects of cell morphology and in variations in metabolic strategies, motility, cell division, developmental biology, etc. The production of extracellular substances and cyanotoxins by cyanobacteria illustrates the diverse nature of their interactions with other organisms (i.e. allelopathy) (Rizvi and Rizvi, 1992).

A molecular approach to the systematics of cyanobacteria may be most fruitful for inferring phylogenetic relationships. Macromolecules, such as nucleic acids and proteins, are copies or translations of genetic information. The methods applied involve direct studies of the relevant macromolecules by sequencing, or indirectly by electrophoresis, hybridisation, or immunological procedures (Wilmotte, 1994). Nucleic acid technologies, especially the polymerase chain reaction (PCR), have advanced to the point that it is feasible to amplify and sequence genes and other conserved regions from a single cell. To date, 16S rRNA has given the most detailed information on the relationships within the cyanobacteria (Rudi *et al.*, 1997). However, the molecular results obtained should be integrated with other characteristics as the base for a polyphasic taxonomy (Vandamme, *et al.*, 1996). A considerable morphological, as well as a genotypical, polymorphy exists in the cyanobacteria, although as data from rRNA sequencing indicates they are correlated to a high degree.

The phylogenetic relationship of cyanobacteria is the rationale behind the meaningful systematic groupings. However, it is difficult to set up a system of classification that serves both the everyday need for practical identification, and offers an expression of the natural relationship between the organisms in question (Mayr, 1981). Meanwhile, it will be necessary to use the available manuals and reference books to help in these investigations and with the proper identification of the cyanobacteria.

Because they are photoautotrophs, cyanobacteria can be grown in simple mineral media. Vitamin B_{12} is the only growth factor that is known to be required by some species. Media must be supplemented with the essential nutrients needed to support cell growth, including sources of nitrogen, phosphorus, trace elements, etc. Toxigenic strains of cyanobacteria are deposited in international-type culture collections (Rippka, 1988; Sugawara *et al.*, 1993). Clonal cultures are distributed for research, taxonomic work and teaching purposes.

Practical Scope

The cyanobacteria have both beneficial and detrimental properties when judged from a human perspective. Their extensive growth can create considerable nuisance for management of inland waters (water supply, recreation, fishing, etc.) and they also release substances into the water which may be unpleasant (Juttner, 1987) or toxic (Gorham and Carmichael, 1988). The water quality problems caused by dense populations of cyanobacteria are intricate, many and various (Skulberg, 1996) and can have great health and economic impacts. As a consequence, the negative aspects of cyanobacteria have gained research attention and public concern.

The properties that make the cyanobacteria generally undesirable are also the qualifications for possible positive economic use. Blue-greens are the source of many valuable products (Richmond, 1990) and carry promising physiological processes, including light-induced hydrogen evolution by biophotolysis (Skulberg, 1994). Extensive research has taken place in the relevant fields of biotechnology.

Cyanobacteria may be used for food or fodder because some strains have a very high content of proteins, vitamins and other essential growth factors and vital pigments of interest can also be produced (Borowitzka and Borowitzka, 1988). Cyanobacteria are also sources for substances of pharmaceutical interest (such as antibiotics) (Falch *et al.*, 1995). These examples are only a few of the possible applications of cyanobacteria for economic development and their exploitation is among the many challenges for biotechnology for the next millennium. Also in this context, their secondary metabolites and health relationships will become important.

Factors Affecting Bloom Formation

Cyanobacteria have a number of special properties which determine their relative importance in phytoplankton communities. However, the behaviour of different cyanobacterial taxa in nature is not homogeneous because their ecophysiological properties differ. An understanding of their response to environmental factors is fundamental for setting water management targets. Because some cyanobacteria show similar ecological and ecophysiological characteristics, they can be grouped by their behaviour in planktonic ecosystems as "ecostrategists" typically inhabiting different niches of aquatic ecosystems. A number of properties and reactions to environmental conditions in order to describe these ecostrategists and to aid the understanding of their specific behaviour.

Light Intensity

Like algae, cyanobacteria contain chlorophyll *a* as a major pigment for harvesting light and conducting photosynthesis. They also contain other pigments such as the phycobiliproteins which include allophycocyanin (blue), phycocyanin (blue) and sometimes phycoerythrine (red) (Cohen-Bazir and Bryant, 1982). These pigments harvest light in the green, yellow and orange part of the spectrum (500-650 nm) which is hardly used by other phytoplankton species. The phycobiliproteins, together with chlorophyll *a*, enable cyanobacteria to harvest light energy efficiently and to live in an environment with only green light. Many cyanobacteria are sensitive to prolonged periods of high light intensities. The growth of *Planktothrix* (formerly *Oscillatoria) agardhii* is inhibited when exposed for extended periods to light intensities above 180 µE m-2 s-1. Long exposures at light intensities of 320 µE m-2 s-1 are lethal for many species (Van Liere and Mur, 1980).

However, if exposed intermittently to this high light intensity, cyanobacteria grow at their approximate maximal rate (Loogman, 1982). This light intensity amounts to less than half of the light intensity at the surface of a lake, which can reach 700-1,000 µE m-2 s-1.

Cyanobacteria which form surface blooms seem to have a higher tolerance for high light intensities. Paerl *et al.* (1983) related this to an increase in carotenoid production which protects the cells from photoinhibition.

Cyanobacteria are further characterised by a favourable energy balance. Their maintenance constant is low which means that they require little energy to maintain cell function and structure (Gons, 1977; Van Liere *et al.*, 1979). As a result of this, the cyanobacteria can maintain a relatively higher growth rate than other phytoplankton organisms when light intensities are low. The cyanobacteria will therefore have a competitive advantage in lakes which are turbid due to dense growths of other phytoplankton.

This was demonstrated in an investigation measuring growth of different species of phytoplankton at various depths in a eutrophic Norwegian lake (Kallqvist, 1981). The results showed that the diatoms *Asterionella, Diatoma* and *Synedra* grew faster than the cyanobacterium *Planktothrix* at 1 m depth, while the growth rate was about the same for all these organisms at 2 m depth. At the very low light intensities below 3 m only *Planktothrix* grew. The ability of cyanobacteria to grow at low light intensities and to harvest certain specific light qualities, enables them to grow in the "shadow" of other phytoplankton. Van Liere and Mur (1979) demonstrated competition between cyanobacteria and other phytoplankton.

Whereas the green alga *(Scenedesmus protuberans)* grew faster at high light intensities, growth of the cyanobacterium *(Planktothrix agardhii)* was faster at low light intensities. If both organisms were grown in the same continuous culture at low light intensity, *Planktothrix* could out-compete *Scenedesmus.* At high light intensities, the biomass of the green alga increased rapidly, causing an increase in turbidity and a decrease in light availability. This increased the growth rate of the cyanobacterium, which then became dominant after 20 days. Although cyanobacteria cannot reach the maximum growth rates of green algae, at very low light intensities their growth rate is higher. Therefore, in waters with high turbidity they have better chances of out-competing other species. This can explain why cyanobacteria which can grow under very poor nutritional conditions often develop blooms in nutrient-rich eutrophic waters.

The light conditions in a given water body determine the extent to which the physiological properties of cyanobacteria will be of advantage in their competition against other phytoplankton organisms (Mur *et al.*, 1978). The zone in which photosynthesis can occur is termed the euphotic zone (Zeu). By definition, the euphotic zone extends from the surface to the depth at which 1 per cent of the surface light intensity can be detected. It can be estimated by measuring transparency with a Secchi disk and multiplying the Secchi depth reading by a factor of 2-3. The euphotic zone may be deeper or more shallow than the mixed, upper zone of a thermally stratified water body, the depth of which is termed the epilimnion (Zm). Many species of planktonic algae and cyanobacteria have little, or only weak, means of active movement and are passively entrained in the water circulation within the epilimnion.

Thus, they can be photosynthetically active only when the circulation maintains them in the euphotic zone. In eutrophic waters, phytoplankton biomass is frequently very high and causes substantial turbidity. In such situations, the euphotic zone is often more shallow than the epilimnion, i.e. the ratio Zeu/Zm is < 1, and phytoplankton spend part of the daylight period in the dark. Thus, the Zeu/Zm ratio is a reasonable (and easy to measure) approach to describing the light conditions encountered by the planktonic organisms.

Gas Vesicles

Many planktonic cyanobacteria contain gas vacuoles (Walsby, 1981). These structures are aggregates of gas-filled vesicles, which are hollow chambers with a hydrophilic outer surface and a hydrophobic inner surface (Walsby, 1978). A gas vesicle has a density of about one tenth that of water (Walsby, 1987) and thus gas vesicles can give cyanobacterial cells a lower density than water.

Growth Rate

The growth rate of cyanobacteria is usually much lower than that of many algal species (Hoogenhout and Amesz, 1965; Reynolds, 1984). At 20°C and light saturation, most common planktonic cyanobacteria achieve growth rates of 0.3-1.4 doublings per day, while diatoms reach 0.8-1.9 doublings per day and growth rates of up to 1.3-2.3 doublings per day have been observed for single-celled green algae (Van Liere and Walsby, 1982). Slow growth rates require long water retention times to enable a bloom of cyanobacteria to form. Therefore cyanobacteria do not bloom in water with short retention times. A. comprehensive overview of mechanisms determining the growth rates of planktonic algae and cyanobacteria under different field conditions is available in Reynolds (1997).

Phosphorus and Nitrogen

Because cyanobacterial blooms often develop in eutrophic lakes, it was originally assumed that they required high phosphorus and nitrogen concentrations. This assumption was maintained even though cyanobacterial blooms often occurred when concentrations of dissolved phosphate were lowest. Experimental data have shown that the affinity of many cyanobacteria for nitrogen or phosphorus is higher than for many other photosynthetic organisms. This means that they can out-compete other phytoplankton organisms under conditions of phosphorus or nitrogen limitation. In addition to their high nutrient affinity, cyanobacteria have a substantial storage capacity for phosphorus.

They can store enough phosphorus to perform two to four cell divisions, which corresponds to a 4-32 fold increase in biomass. However, if total phosphate rather than only dissolved phosphate is considered, high concentrations indirectly support cyanobacteria because they provide a high carrying capacity for phytoplankton. High phytoplankton density leads to high turbidity and low light availability, and cyanobacteria are the group of phytoplankton organisms which can grow best under these conditions.

A low ratio between nitrogen and phosphorus concentrations may favour the development of cyanobacterial blooms. A comparison between the optimum N:P ratios for eukaryotic algae (16-23 molecules N:1 molecule of P) with the optimum rates for bloom-forming cyanobacteria (10-16 molecules N: 1 molecule P), shows that the ratio is lower for cyanobacteria (Schreurs, 1992).

Population Stability

While many planktonic algae are grazed by copepods, daphnids and protozoa, cyanobacteria are not grazed to the same extent, and the impact of grazing by some specialised ciliates and rhizopod protozoans is usually not substantial.

Cyanobacteria are attacked by viruses, bacteria and actinomycetes, but the importance of these natural enemies for the breakdown of populations is not well understood. Because they have few natural enemies, and their capacity for buoyancy regulation prevents sedimentation, the loss rates of cyanobacterial populations are generally low. Thus, their slow growth rates are compensated by the high prevalence of populations once they have been established.

Temperature

Maximum growth rates are attained by most cyanobacteria at temperatures above 25°C (Robarts and Zohary, 1987). These optimum temperatures are higher than for green algae and diatoms. This can explain why in temperate and boreal water bodies most cyanobacteria bloom during summer.

Cyanobacterial Ecostrategists

The physiological properties of cyanobacteria discussed above vary between different species. As a consequence, different "ecostrategists" inhabit different types of water bodies. A preliminary approach to describing these ecostrategists, based on ecophysiological laboratory work together with field observations (largely from northwestern Europe). This information may be useful for management, because it helps to predict which cyanobacteria can be expected to occur under certain conditions. Further development of this approach will be possible as more data on occurrence of cyanobacteria under different growth conditions are collected from other continents.

Scum-forming Ecostrategists

During the vegetation period, a number of cyanobacteria develop large aggregates (colonies) of coccoid cells or filaments which are not homogeneously distributed over the water column. Important genera showing this development are *Microcystis, Anabaena* and *Aphanizomenon.* At the water surface the rate of photosynthesis of the colonies is high and the cells store large quantities of carbohydrates. Although the cells contain gas vesicles, the heavy carbohydrates acts as ballast and induce sinking within the colonies.

According to Stoke's Law the sinking rate is dependent on the difference in density between the water and the cells, and on the square of the colony size (d_2). Large colonies sink faster than small ones, and single cells hardly show any vertical migration. By sinking, colonies move out of the euphotic zone into the deeper, dark water layers, where they use their carbohydrates during respiration and synthesise new gas-vesicles (Utkilen *et al.*, 1985). They then become buoyant again and return to the euphotic zone.

Buoyancy regulation enables the colonies to position themselves in light conditions which are optimal for their growth. A prerequisite is that the water body is not too turbulent. During the night, all colonies may become buoyant and some of the population may be accumulated on the water surface where they can be blown together by wind, forming stable scums along downwind shores. The frequency of vertical migration is dependent upon colony size. In temperate regions, as temperatures decline in the autumn, photosynthesis becomes more rapid than respiration, and the carbohydrate "ballast" is not consumed. The colonies therefore sink to the bottom of the water body where they may survive the winter, gradually consuming their carbohydrate stores by respiration or fermentation. Cells which re-ascend from the bottom in the spring are unicellular or formed into very small colonies. During this period *Microcystis* spp. is difficult to recognise in plankton samples, and only becomes more conspicuous when the colonies increase in size during early summer.

Buoyancy regulation can be a substantial advantage in competition with other phytoplankton organisms. However, this type of regulation is only possible in water bodies with a shallow euphotic zone in relation to the depth of vertical mixing (Zeu < Zm). Therefore, in temperate climates, blooms of *Microcystis* spp. are found particularly in water bodies deeper than 3 m, because the euphotic zone is likely to be substantially more shallow than the mixed depth. However, even in shallow lakes, where they do not have the competitive advantage of vertical migration, *Microcystis* spp. may become dominant and form substantial blooms, as has been reported from Hungary, Australia, and particularly from subtropical and tropical regions. Reynolds (1997) characterises

Microcystis spp. as notoriously and overwhelmingly dominant in some lakes of the lower latitudes that exhibit diel stratification.

Many cyanobacteria cannot survive high light intensities over longer periods. This may limit their distribution to more turbid, eutrophic ecosystems. However, *Microcystis* species are less sensitive to high light intensities because buoyancy regulation enables them to find light conditions that are optimal for their growth. This means that the presence of *Microcystis* cannot be related strictly to the level of eutrophication. The genus is therefore found in mesotrophic, eutrophic and in hypertrophic waters. However, the amount of biomass that this species can attain depends on the level of eutrophication. Most *Microcystis* blooms are found in lakes with an average summer chlorophyll *a* concentration of 20-50 µg l-1 and a Secchi transparency of 1-2 m.

Homogeneously Dispersed Ecostrategists

This ecotype comprises filamentous species, such as *Planktothrix (Oscillatoria) agardhii* and *Limnothrix (Oscillatoria) redekei*. These species are extremely sensitive to high light intensities and do not form colonies (Reynolds, 1987). Because the filaments are quite small, vertical migration by buoyancy regulation is less pronounced than their passive entrainment by water circulation. Therefore, these species are homogeneously dispersed throughout the epilimnion.

This type of ecostrategist is found in eutrophic and hypertrophic shallow lakes. Many lakes with blooms of dispersed ecotypes have a depth of not more than 3 m and chlorophyll concentrations of 50 µg l-1 and, in extreme cases, greater than 200 µg l-1. The filaments are hardly grazed and do not sediment. Blooms of this type often lead to virtual monocultures which can prevail year-round for many years. Population dynamics in such lakes can be limited. In temperate regions, the autumn population can even survive under ice in winter. In such situations, the spring population starts growth with a relatively high density and thus has an advantage in competition with other species (Visser, 1990). By causing high turbidity, these cyanobacterial populations effectively suppress the growth of other phytoplankton species. Thus, the next summer population establishes itself almost without any seasonal succession between different species of phytoplankton. This high stability of the population precludes any redistribution of phosphorus and nitrogen to other components of the ecosystem and this can cause a resilience effect in lake restoration projects.

Stratifying Ecostrategists

Representatives of this ecotype develop stable summer populations in the intermediate zone of thermally stratified lakes and reservoirs known as the metalimnion see. The organisms contain the red pigment phycoerythrin to absorb the green light, which is the prevailing wave length at this depth. The most common of these species is *Planktothrix (Oscillatoria) rubescens,* but red varieties of other *Planktothrix* species can also form metalimnic populations (Aune *et al.*, 1997).

The single filaments of these species hardly show any vertical migration. However, in late autumn at the end of the growing season, the cells can become buoyant and then form red surface scums (Walsby *et al.*, 1983). The niche of this type of *Planktothrix* is very limited. It needs sufficient light in the metalimnetic zone, but may be inhibited by too much light. Most metalimnetic blooms are found at light intensities of 1-5 per cent of the surface irradiance and in a range of Zeu/Zm between 0.7 and 1.2.

Nitrogen Fixing Ecostrategists

The mass development of species capable of fixation of atmospheric nitrogen (species of the genera *Anabaena, Aphanizomenon, Cylindrospermopsis, Nodularia,* and *Nostoc)* can often be related to periodic nitrogen limitation. Examples are found in deep, as well as in shallow, systems. However, while these ecostrategists often dominate in ecosystems with low levels of inorganic dissolved nitrogen, the reverse does not necessarily apply. Numerous lakes with clear nitrogen limitation are not dominated by nitrogen-fixing cyanobacteria. Low light availability may be the reason for this, because nitrogen fixation requires high amounts of energy. In turbid lakes, insufficient light energy may be available for effective nitrogen fixation (Zevenboom and Mur, 1980). A number of nitrogen fixing species can form colonies and possess gas vesicles. This means that they can regulate buoyancy, like *Microcystis,* and can form stable scums along downwind shores.

Restoration measures which simultaneously reduce phosphate and nitrogen loading (sewage diversion, isolation) may strengthen prevailing nitrogen-limiting conditions and hence the probability of large populations of nitrogen-fixing cyanobacteria.

Small, colony-forming taxa Cases of large populations of the small, colony-forming genus *Aphanothece* have been reported. Little information is available on buoyancy regulation and scum formation by the species involved. In several water bodies, *Aphanothece* dominance has occurred after a decrease of *Planktothrix rubescens* populations. Reynolds (1997) reported them as the only cyanobacteria present in the summer plankton of small, intermittently flushed lakes in England. The dominance of this group is not strictly related with phosphate or nitrogen limitation, and there are no obvious relationships that can explain the sudden dominance of these cyanobacteria. They seem to dominate in an intermediate state during lake recovery after restoration measures have been taken; their ecology is unknown.

Benthic Cyanobacteria

Besides the planktonic ecostrategists, cyanobacteria may grow on the bottom sediments of water bodies which are sufficiently clear to allow light penetration to these surfaces. These benthic species may form coherent mats. Especially high rates of photosynthesis by such mats sometimes leads to trapping of the photosynthetically produced oxygen as bubbles within the mats; parts of the mats may then become sufficiently buoyant to tear loose and rise to the surface. For monitoring and management of toxic cyanobacteria, awareness of these is important because cyanotoxin problems are usually not expected in clear, oligotrophic waters.

However, toxic benthic cyanobacteria have caused animal deaths in Scotland, where beached mats along the shore of a clear loch were scavenged by dogs (Gunn *et al.*, 1992), and in Switzerland where toxic benthic populations of *Oscillatoria limosa* were ingested by cattle drinking from pristine mountain lakes (Metz *et al.*, 1997, 1998).

Toxin Production by Cyanobacteria

The worldwide occurrence of toxic cyanobacterial blooms in fresh and brackish eutrophic waters creates a problem for all life forms. Most water-based poisonings by cyanobacteria occur when heavy surface growths or scums accumulate near shorelines of lakes, ponds and reservoirs where animals have free access to high concentrations of these toxic cells. Deaths attributed to cyanobacterial toxins have been reported for man, animals, birds and fish.

Cyanobacterial blooms are ubiquitous, often associated with eutrophication and appear to be on the increase, also in South Africa. N, P and C are important nutrients for high growth rates and the ratios in the supply concentrations are often decisive in selecting for cyanobacterial dominance. These organisms are capable of scavenging their environments for resources and excessive or "luxury" uptake of nutrients allow them to survive extreme nutrient deficient conditions. Cyanobacteria flourish at high temperatures, neutral to alkaline conditions, high nutrient concentrations especially where the ratios of N:P are low and an adequate supply of iron is present. The buoyancy of certain species, due to especially the production of gas-vacuoles gives them a competitive advantage over other phytoplankton. All blooms of *Microcystis*, one of the most ubiquitous species, should be considered to be toxic and toxins are easily leached from the cells. They persist for long periods and are not easily degraded.

Microcystis aeruginosa is distributed worldwide that often form seasonal blooms, as is common in South Africa. This organism produces a vast number of peptides (microcystins), some of which are highly toxic. Microcystins consist of a seven-membered peptide ring made up of five non-protein amino acids and two protein amino acids. More than 50 microcystins have been identified to date, most representing minor components of the total toxin complement of the cyanobacteria from which they were isolated. The most commonly occurring toxin is microcystin- LR, a cyclic heptapeptide hepatotoxin, where symptoms of exposure too, includes skin irritation, possible liver cancer as a result of cronic exposure, and even death.

These toxins are inhibitors of serine/threonine protein phosphatase enzymes and are among the most potent tumour-promoting compounds. The mechanism of toxicity is exerted by the general inhibition of dephosphorylation of protein phosphatases 1 and 2A, leading to hyperphosphorylation in the cytosol.

The genus *Microcystis* contains non-toxic and toxic strains and the toxicity is affected by various environmental factors such as water temperature, pH, intensity of solar radiation, dissolved oxygen and CO_2 availability. It has been reported that microcystin concentrations are usually higher under conditions, when the organisms are stressed by any of the above factors.

The molecular basis of toxin-production in *M. aeruginosa* was partially elucidated elsewhere and it was found that both toxin-producing and non- toxin-producing strains of *M. aeruginosa* contained sequences that revealed a high degree of homology with several well-characterised peptide synthetases. In blotting experiments, a PCR fragment based on a portion of one of these peptide synthetases hybridised exclusively to restricted genomic DNA from toxin-producing strains indicating that this peptide synthetase was involved in toxin production.

9

Element Cycling

The biotransformation is a procedure in the metabolism of organisms, with which separatable materials do not become converted into separatable materials by chemical processes (*transformed*).

During the physiological metabolism of the body substances result again and again, which cannot be separated directly over the urine or the faeces. Usually these materials (Gallenfarbstoffe, Steroidhormone) are not at allwater soluble lipophilically. Beyond that the organism takes foreign matter from nature with the food on (Xenobiotika), or from humans synthesised substances (main medicines, drugs, preservative, pestizide, etc.).

An accumulation (accumulation) of these substances in the body would be fatal and finally deadly. Around these substances into a separatable form to transfer many fabrics, above all however the liver, are capable of the biotransformation.

The biotransformation can be divided in 2 phases (Phase I and Phase II). Phase I (transformation reactions) insert functional groups (-OH; SH) into the non-polar molecules. In the Phase II (conjugation reactions) the molecules are connected by the functional groups with water-soluble molecules (conjugated) and to be able either by the kidneys or by the Galle to be then separated.

Phase Irish Actions [work on]

- Hydrolytic splitting (z. B. acetylsalicylic acid (aspirin) is amplifier-off-changed over this way.
- Methylierungen (for example the Catechyl-o-methyl. Catechyl-o-methyl-transferase converts for Catechyle (Noradrenalin, adrenalin, Dopamine).
- Reduction reactions
- Oxidation reactions (Cytochrom P450 dependently)
  - Aromatic Hydroxylierung
  - Aliphatic Hydroxylierung
  - Epoxy formation
  - Dealkylierung

Phase II-reactions

In the second phase the foreign matter or metabolites of the first phase is connected with endogenous, usually strongly water-soluble materials. To the reactions of the Phase II belong:

- Glucoronidierung with Glucoronat (GluUA). This reaction is the most important. With their assistance Steroide, bilirubin and phenol rings are eliminated.
- Veresterung with sulphate.
- Conjugation with amino acids.

All reactions of the Phase II are taken over by transferasen.

Cytochrom

These with *CYP 450* shortened, mikrosomalen (in the smooth endoplasmatischen Retikulum special meaning comes) to enzymes present with the biotransformation. Their task in the Phase I reaction consists of transferring molecular oxygen into a reactive form and building these into the substrates. They need for it the Coenzyme NADPH and Flavin Adenine Dinukleotid (FAD).

According to the multiplicity of poison materials a whole pallet of CYP 450 exists. They exhibit a small substrate specificity, which means that an enzyme can convert many materials (substrates). The Expression of these enzymes is initiated by their substrates. Like that it is understandable that medicines, which are diminished over the same CYP 450 affect each other mutually in their Pharmakokinetik. From the daily life well-known examples of it are:

- Contents materials of the grapefruit juice inactivate CYP 3A4, while them are induced by preparing free-for sale with Johanniskraut (when mild anti-depressive). This enzyme diminishes also a large number of medicines, e.g. with blood high pressure (hypertonia) to be used up (Verapamil, Nifedipin); in addition, Antikonvulsiva such as Phenytoin and diazepam or anti-arrhythmics such as Amiodaron as well as many different.
- Alcohol (ethanol) is diminished over the same *microsomale ethanol-oxidising system (MEOS)* as methanol or some antibiotics. Those leads to the fact that the effects of these medicines can strengthen themselves with consumption of alcohol dangerously.
- Antibiotics such as Rifampicin induce the strengthened Expression of different CYP 450. Thus active substances of the Kontrazeptiva (pill) are diminished faster and lose their effect.

Prodrugs and Giftung

Some substances are taken up in an inactive form. Only by the transformation by the biotransformation they will transfer into the actual effective form. If it concerns thereby medicines, like z. B. the sleeping drug Chlordiazepoxid, then one calls these medicines Prodrugs. If the supplied substance is transferred by the biotransformation in poisonous metabolites, one calls this procedure Giftung. Best example of it is methanol, which is alone innocuous, however in the dismantling way into the highly poisonous formaldehyde and into formic acid is later converted. Similarly as the Prodrugs morphine is transferred by Glucuronidierung in the liver into the Morphin-6-Glucuronid, which works still substantially more strongly than morphine. These effects are called also roofridge passport effect.

Nitrogen Cycle

The nitrogen cycle is the biogeochemical cycle that describes the transformations of nitrogen and nitrogen-containing compounds in nature.

Earth's atmosphere is about 78 per cent nitrogen, making it the largest pool of nitrogen. Nitrogen is essential for many biological processes; and is crucial for any life here on Earth. It is in all amino acids, is incorporated into proteins, and is present in the bases that make up nucleic acids, such as DNA and RNA. In plants, much of the nitrogen is used in chlorophyll molecules which are essential for photosynthesis and further growth.

Processing, or fixation, is necessary to convert gaseous nitrogen into forms usable by living organisms. Some fixation occurs in lightning strikes, but most fixation is done by free-living or symbiotic bacteria. These bacteria have the nitrogenase enzyme that combines gaseous nitrogen with hydrogen to produce ammonia, which is then further converted by the bacteria to make its own organic compounds. Some nitrogen fixing bacteria, such as *Rhizobium*, live in the root nodules of legumes (such as peas or beans). Here they form a mutualistic relationship with the plant, producing ammonia in exchange for carbohydrates. Nutrient-poor soils can be planted with legumes to enrich them with nitrogen. A few other plants can form such symbioses.

Other plants get nitrogen from the soil by absorption at their roots in the form of either nitrate ions or ammonium ions. All nitrogen obtained by animals can be traced back to the eating of plants at some stage of the food chain.

Ammonia

The source of ammonia is the decomposition of dead organic matter by bacteria called *decomposers,* which produce ammonium ions (NH_4^+). In well-oxygenated soil, these ions are then oxygenated first by nitrifying bacteria into nitrite (NO_2^-) and then into nitrate (NO_3^-). This two-step conversion of ammonium into nitrate is called *nitrification.*

Ammonium ions readily bind to soils, especially to humic substances and clays. Nitrate and nitrite ions, due to their negative electric charge, bind less readily since there are less positively charged ion-exchange sites (mostly humic substances) in soil than negative. After rain or irrigation, *leaching* (the removal of soluble ions, such as nitrate and nitrite) into groundwater can occur. Elevated nitrate in groundwater is a concern for drinking water use because nitrate can interfere with blood-oxygen levels in infants and cause methemoglobinemia or blue-baby syndrome.

Where groundwater recharges stream flow, nitrate-enriched groundwater can contribute to eutrophication, a process leading to high algal, especially blue-green algal populations and the death of aquatic life due to excessive demand for oxygen. While not directly toxic to fish life like ammonia, nitrate can have indirect effects on fish if it contributes to this eutrophication. Nitrogen has contributed to severe eutrophication problems in some water bodies. As of 2006, the application of nitrogen fertiliser is being increasingly controlled in Britain and the United States. This is occurring along the same lines as control of phosphorus fertilizer, restriction of which is normally considered essential to the recovery of eutrophied waterbodies.

Ammonia is highly toxic to fish life and the water discharge level of ammonia from wastewater treatment plants must often be closely monitored. To prevent loss of fish, nitrification prior to discharge is often desirable. Land application can be an attractive alternative to the mechanical aeration needed for nitrification.

During anaerobic (low oxygen) conditions, *denitrification* by bacteria occurs. This results in nitrates being converted to nitrogen gas and returned to the atmosphere. Nitrate can also be reduced to nitrite and subsequently combine with ammonium in the anammox process, which also results in the production of dinitrogen gas.

Processes of the Nitrogen Cycle

Nitrogen fixation

The conversion of dinitrogen (N_2) from the atmosphere into a form readily available to plants and hence to animals and humans is an important step in the nitrogen cycle, that determines the supply of this essential nutrient. There are four ways to convert N_2 (atmospheric nitrogen gas) into more chemically reactive forms:

1. *Biological Fixation:* Some symbiotic bacteria (most often associated with leguminous plants) and some free-living bacteria are able to fix nitrogen and assimilate it as organic nitrogen. An example of mutualistic nitrogen fixing bacteria are the *Rhizobium* bacteria, which live in legume root nodules. These species are diazotrophs. An example of the free-living bacteria is *Azotobacter*.
2. *Industrial N-fixation:* In the Haber-Bosch process, N_2 is converted together with hydrogen gas (H_2) into ammonia (NH_3) fertilizer and explosives.
3. *Combustion of Fossil Fuels:* Automobile engines and thermal power plants, which release NOx.
4. *Other Processes:* Additionally, the formation of NO from N_2 and O_2 due to photons and especially lightning, are

important for atmospheric chemistry, but not for terrestrial or aquatic nitrogen turnover.

Assimilation

Plants can absorb nitrate or ammonium ions from the soil via their root hairs. If nitrate is absorbed, it is first reduced to nitrite ions and then ammonium ions for incorporation into amino acids, nucleic acids, and chlorophyll. In plants which have a mutualistic relationship with rhizobia, some nitrogen is assimilated in the form of ammonium ions directly from the nodules. Animals, fungi, and other heterotrophic organisms absorb nitrogen as amino acids, nucleotides and other small organic molecules.

Ammonification

When a plant or animal dies, or an animal excretes, the initial form of nitrogen is organic. Bacteria, or in some cases, fungi, convert the organic nitrogen within the remains back into ammonia, a process called ammonification or mineralisation.

Nitrification

The conversion of ammonia to nitrates is performed primarily by soil-living bacteria and other nitrifying bacteria. The primary stage of nitrification, the oxidation of ammonia (NH_3) is performed by bacteria such as the *Nitrosomonas* species, which converts ammonia to nitrites (NO_2^-). Other bacterial species, such as the *Nitrobacter*, are responsible for the oxidation of the nitrites into nitrates (NO_3^-).

Denitrification

Denitrification is the reduction of nitrites back into the largely inert nitrogen gas (N_2), completing the nitrogen cycle. This process is performed by bacterial species such as the *Pseudomonas* and *Clostridium* in anaerobic conditions. They use the nitrate as an electron acceptor in the place of oxygen. These facultatively anaerobic bacteria can also live in aerobic conditions.

Anaerobic Ammonium Oxidation

In this biological process, nitrite and ammonium are converted directly into dinitrogen gas. This process makes up a major proportion of dinitrogen conversion in the oceans.

Human Influences on the Nitrogen Cycle

As a result of extensive cultivation of legumes (particularly soy, alfalfa, and clover), growing use of the Haber-Bosch process in the creation of chemical fertilizers, and pollution emitted by vehicles and industrial plants, human beings have more than doubled the annual transfer of nitrogen into biologically available forms. In addition, humans have significantly contributed to the transfer of nitrogen trace gases from Earth to the atmosphere, and from the land to aquatic systems.

N_2O has risen in the atmosphere as a result of agricultural fertilization, biomass burning, cattle and feedlots, and other industrial sources. N_2O has deleterious effects in the stratosphere, where it breaks down and acts as a catalyst in the destruction of atmospheric ozone. Ammonia (NH_3) in the atmosphere has tripled as the result of human activities. It is a reactant in the atmosphere, where it acts as an aerosol, decreasing air quality and clinging on to water droplets, eventually resulting in acid rain. Fossil fuel combustion has contributed to a 6 or 7 fold increase in NOx flux to the atmosphere. NOx actively alters atmospheric chemistry, and is a precursor of tropospheric (lower atmosphere) ozone production, which contributes to smog, acid rain, and increases nitrogen inputs to ecosystems. Ecosystem processes can increase with nitrogen fertilization, but anthropogenic input can also result in nitrogen saturation, which weakens productivity and can kill plants. Decreases in biodiversity can also result if higher nitrogen availability increases nitrogen-demanding grasses, causing a degradation of nitrogen-poor, species diverse heathlands.

Oxygen Cycle

The oxygen cycle is the biogeochemical cycle that describes the movement of oxygen within and between its three main reservoirs: the atmosphere, the biosphere, and the lithosphere. The main driving factor of the oxygen cycle is photosynthesis, which is responsible for the modern Earth's atmosphere and life as we know it. Because of the vast amounts of oxygen in the atmosphere, even if all photosynthesis were to cease it would take between 5,000 to 2.5 million years to strip out more or less all oxygen.

Reservoirs and Fluxes

By far the largest reservoir of Earth's oxygen is within the silicate and oxide minerals of the crust and mantle (99.5 per cent). Only a small fraction has been released as free oxygen to the biosphere (0.01 per cent) and atmosphere (0.49 per cent).

The main source of oxygen within the biosphere and atmosphere is photosynthesis, which breaks down carbon dioxide and water to create sugars and oxygen:

$$6CO_2 + 6H_2O + \text{energy} \rightarrow C_6H_{12}O_6 + 6O_2$$

Photosynthesising organisms include the plant life of the land areas as well as the phytoplankton of the oceans. The tiny marine cyanobacteria Prochlorococcus was discovered in 1986 and accounts for more than half of the photosynthesis of the open ocean.

An additional source of atmospheric oxygen comes from photolysis, whereby high energy ultraviolet radiation breaks down atmospheric water and nitrite into component atoms. The free H and N atoms escape into space leaving O_2 in the atmosphere:

$$2H_2O + \text{energy} \rightarrow 4H + O_2$$

$$2N_2O + \text{energy} \rightarrow 4N + O_2$$

The main way oxygen is lost from the atmosphere is via respiration and decay mechanisms in which animal life and bacteria consume oxygen and release carbon dioxide.

Because lithospheric minerals are oxidised in oxygen, chemical weathering of exposed rocks also consumes oxygen.

An example of surface weathering chemistry is formation of iron-oxides (rust):

$$4Fe + 3O_2 \rightarrow 2Fe_2O_3$$

Mineral Redox Buffer

Oxygen is also cycled between the biosphere and lithosphere. Marine organisms in the biosphere create calcium carbonate shell material ($CaCO_3$) that is rich in oxygen. When the organism dies its shell is deposited on the shallow sea floor and buried over time to create the limestone rock of the lithosphere. Weathering processes initiated by organisms can also free oxygen from the lithosphere. Plants and animals extract nutrient minerals from rocks and release oxygen in the process.

Oxygen Reservoir Capacities and Fluxes

The following tables offer estimates of oxygen cycle reservoir capacities and fluxes. These numbers are based primarily on estimates from (Walker, J.C.G.):

Table: Major reservoirs involved in the oxygen cycle

Reservoir	*Capacity (kg O_2)*	*Flux In/Out (kg O_2 per year)*	*Residence Time (years)*
Atmosphere	$1.4 * 10^{18}$	$30,000 * 10^{10}$	4,500
Biosphere	$1.6 * 10^{16}$	$30,000 * 10^{10}$	50
Lithosphere	$2.9 * 10^{20}$	$60 * 10^{10}$	500,000,000

Table: Annual gain and loss of atmospheric oxygen (Units of 10^{10} kg O_2 per year)

Gains	
Photosynthesis (land)	16,500
Photosynthesis (ocean)	13,500
Photolysis of N_2O	1.3
Photolysis of H_2O	0.03
	Total Gains ~ 30,000

Contd...

Gains	
Losses - Respiration and Decay	
Aerobic Respiration	23,000
Microbial Oxidation	5,100
Combustion of Fossil Fuel (anthropologic)	1,200
Photochemical Oxidation	600
Fixation of N_2 by Lightning	12
Fixation of N_2 by Industry (anthropologic)	10
Oxidation of Volcanic Gases	5
Losses - Weathering	
Chemical Weathering	50
Surface Reaction of O_3	12
Total Losses ~ 30,000	

Ozone

The presence of atmospheric oxygen has led to the formation of ozone and the ozone layer within the stratosphere. The ozone layer is extremely important to modern life as it absorbs harmful ultraviolet radiation:

$$O_2 + \text{uv energy} \rightarrow 2O$$

$$O + O_2 \rightarrow O_3$$

The absorbed solar energy also raises the temperature of the atmosphere within the ozone layer creating a thermal barrier that helps trap the atmosphere below (as opposed to bleeding out into space).

Phosphorus

An interesting theory is that phosphorus (P) in the ocean helps regulate the amount of atmospheric oxygen. Phosphorus dissolved in the oceans is an essential nutrient to photosynthetic life and one of the key limiting factors. Oceanic photosynthesis contributes approximately 45 per cent of the total free oxygen to the oxygen cycle. The population growth of photosynthetic organisms is primarily limited by the availability of dissolved phosphorus.

One side-effect of mining and industrial activities is a dramatic increase in the amount of phosphorus being discharged to the world's oceans. However, this increase in available phosphorus has not resulted in a corresponding increase in oceanic photosynthesis. This is because an increase in photosynthesiser population results in increased oxygen levels in the oceans. The elevated oxygen levels promote the growth of certain types of bacteria that compete for uptake of dissolved phosphorus. This competition limits the amount of phosphorus available to photosynthetic life thus buffering their total population as well as the levels of O_2.

Carbon Cycle

The carbon cycle is the biogeochemical cycle by which carbon is exchanged between the biosphere, geosphere, hydrosphere, and atmosphere of the Earth (other astronomical objects may have similar carbon cycles, but nothing is yet known about them).

The cycle is usually thought of as four major reservoirs of carbon interconnected by pathways of exchange. The reservoirs are the atmosphere, the terrestrial biosphere (which usually includes freshwater systems and non-living organic material, such as soil carbon), the oceans (which includes dissolved inorganic carbon and living and non-living marine biota), and the sediments (which includes fossil fuels). The annual movements of carbon, the carbon exchanges between reservoirs, occur because of various chemical, physical, geological, and biological processes. The ocean contains the largest active pool of carbon near the surface of the Earth, but the deep ocean part of this pool does not rapidly exchange with the atmosphere.

The global carbon budget is the balance of the exchanges (incomes and losses) of carbon between the carbon reservoirs or between one specific loop (e.g., atmosphere - biosphere) of the carbon cycle. An examination of the carbon budget of a pool or reservoir can provide information about whether the pool or reservoir is functioning as a source or sink for carbon dioxide.

In the Atmosphere: Carbon exists in the Earth's atmosphere primarily as the gas carbon dioxide (CO_2). Although it is a very small part of the atmosphere overall (approximately 0.04 per cent on a molar basis, though rising), it plays an important role in supporting life. Other gases containing carbon in the atmosphere are methane and chlorofluorocarbons (the latter is entirely artificial). The overall atmospheric concentration of these greenhouse gases has been increasing in recent decades, contributing to global warming.

Carbon is taken from the Atmosphere in Several Ways:

- When the sun is shining, plants perform photosynthesis to convert carbon dioxide into carbohydrates, releasing oxygen in the process. This process is most prolific in relatively new forests where tree growth is still rapid.
- At the surface of the oceans towards the poles, seawater becomes cooler and more carbonic acid is formed as CO_2 becomes more soluble. This is coupled to the ocean's thermohaline circulation which transports dense surface water into the ocean's interior.
- In upper ocean areas of high biological productivity, organisms convert reduced carbon to tissues, or carbonates to hard body parts such as shells and tests. These are, respectively, oxidised (soft-tissue pump) and redissolved (carbonate pump) at lower average levels of the ocean than those at which they formed, resulting in a downward flow of carbon.
- The weathering of silicate rock. Carbonic acid reacts with weathered rock to produce bicarbonate ions. The bicarbonate ions produced are carried to the ocean, where they are used to make marine carbonates. Unlike dissolved CO_2 in equilibrium or tissues which decay,

weathering does not move the carbon into a reservoir from which it can readily return to the atmosphere.

Carbon can be released back into the atmosphere in many different ways,

- Through the respiration performed by plants and animals. This is an exothermic reaction and it involves the breaking down of glucose (or other organic molecules) into carbon dioxide and water.
- Through the decay of animal and plant matter. Fungi and bacteria break down the carbon compounds in dead animals and plants and convert the carbon to carbon dioxide if oxygen is present, or methane if not.
- Through combustion of organic material which oxidises the carbon it contains, producing carbon dioxide (and other things, like water vapour). Burning fossil fuels such as coal, petroleum products, and natural gas releases carbon that has been stored in the geosphere for millions of years.
- Production of cement. Carbon dioxide is released when limestone (calcium carbonate) is heated to produce lime (calcium oxide), a component of cement.
- At the surface of the oceans where the water becomes warmer, dissolved carbon dioxide is released back into the atmosphere.
- Volcanic eruptions and metamorphism release gases into the atmosphere. These gases include water vapour, carbon dioxide and sulphur dioxide. The carbon dioxide released is roughly equal to the amount removed by silicate weathering; so the two processes, which are the chemical reverse of each other, sum to roughly zero, and do not affect the level of atmospheric carbon dioxide on time scales of less than about 100,000 yr.

In the Biosphere: Around 1900 gigatons of carbon are present in the biosphere. Carbon is an essential part of life on Earth. It plays an important role in the structure, biochemistry, and nutrition of all living cells. And life plays an important role in the carbon cycle:

- Autotrophs are organisms that produce their own organic compounds using carbon dioxide from the air or water in which they live. To do this they require an external source of energy. Almost all autotrophs use solar radiation to provide this, and their production process is called photosynthesis. A small number of autotrophs exploit chemical energy sources in a process called chemosynthesis. The most important autotrophs for the carbon cycle are trees in forests on land and phytoplankton in the Earth's oceans. Photosynthesis follows the reaction $6CO_2 + 6H_2O \rightarrow C_6H_{12}O_6 + 6O_2$.
- Carbon is transferred within the biosphere as heterotrophs feed on other organisms or their parts (e.g., fruits). This includes the uptake of dead organic material (detritus) by fungi and bacteria for fermentation or decay.
- Most carbon leaves the biosphere through respiration. When oxygen is present, aerobic respiration occurs, which releases carbon dioxide into the surrounding air or water, following the reaction $C_6H_{12}O_6 + 6O_2 \rightarrow 6CO_2 + 6H_2O$. Otherwise, anaerobic respiration occurs and releases methane into the surrounding environment, which eventually makes its way into the atmosphere or hydrosphere (e.g., as marsh gas or flatulence).
- Burning of biomass (e.g. forest fires, wood used for heating, anything else organic) can also transfer substantial amounts of carbon to the atmosphere.
- Carbon may also be circulated within the biosphere when dead organic matter (such as peat) becomes incorporated in the geosphere. Animal shells of calcium carbonate, in particular, may eventually become limestone through the process of sedimentation.
- Over-fishing will reduce the amount of Marine Biota

in the sea, and thus decrease the amount of Carbon taken out of the atmosphere by sea creatures, and thus be a direct cause of increasing atmospheric Carbon Dioxide levels, and consequent global warming.

- Much remains to be learned about the cycling of carbon in the deep ocean. For example, a recent discovery is that larvacean mucus houses (commonly known as "sinkers") are created in such large numbers that they can deliver as much carbon to the deep ocean as has been previously detected by sediment traps. Because of their size and composition, these houses are rarely collected in such traps, so most biogeochemical analyses have erroneously ignored them.

Carbon storage in the biosphere is influenced by a number of processes on different time-scales: while net primary productivity follows a diurnal and seasonal cycle, carbon can be stored up to several hundreds of years in trees and up to thousands of years in soils. Changes in those long term carbon pools (e.g. through de- or afforestation or through temperature-related changes in soil respiration) will thus directly affect global warming.

In the Oceans: The seas contain around 36000 gigatonnes of carbon, mostly in the form of bicarbonate ion. Inorganic carbon, that is carbon compounds with no carbon-carbon or carbon-hydrogen bonds, is important in its reactions within water. This carbon exchange becomes important in controlling pH in the ocean and can also vary as a source or sink for carbon. Carbon is readily exchanged between the atmosphere and ocean. In regions of oceanic upwelling, carbon is released to the atmosphere. Conversely, regions of downwelling transfer carbon (CO_2) from the atmosphere to the ocean. When CO_2 enters the ocean, carbonic acid is formed:

$$CO_2 + H_2O \rightarrow H_2CO_3$$

This reaction has a forward and reverse rate, that is, it achieves a chemical equilibrium. Another reaction important in controlling oceanic pH levels is the release of hydrogen ions and bicarbonate. This reaction controls large changes in pH:

$$H_2CO_3 \rightarrow H^+ + HCO_3^-$$

Carbon Cycle Modelling

Models of the carbon cycle can be incorporated into global climate models, so that the interactive response of the oceans and biosphere on future CO_2 levels can be modelled. There are considerable uncertainties in this, both in the physical and biogeochemical submodels (especially the latter). Such models typically show that there is a positive feedback between temperature and CO_2. For example, Zeng *et al.* (*GRL*, 2004) find that in their model, including a coupled carbon cycle increases atmospheric CO_2 by about 90 ppmv at 2100 (over that predicted in models with non-interactive carbon cycles), leading to an extra 0.6°C of warming (which, in turn, may lead to even greater atmospheric CO_2).

Phosphorus Cycle

The phosphorus cycle is the biogeochemical cycle that describes the movement of phosphorus through the lithosphere, hydrosphere, and biosphere. Unlike many other biogeochemicals, the atmosphere does not play a significant role in the movements of phosphorus, because phosphorus and phosphorus-based compounds are usually solids at the typical ranges of temperature and pressure found on Earth.

Phosphorus in the Environment: Phosphorus normally occurs in nature as part of a phosphate ion, consisting of a phosphorus atom and some number of oxygen atoms, the most abundant form (called *orthophosphate*) having four oxygens: PO_4^{3-}. Most phosphates are found as salts in ocean sediments or in rocks. Over time, geologic processes can bring ocean sediments to land, and weathering will carry terrestrial phosphates back to the ocean. Plants absorb phosphates from the soil. The plants may then be consumed by herbivores who in turn may be consumed by carnivores. After death, the animal or plant decays, and the phosphates are returned to the soil. Run-off may carry them back to the ocean or they may be reincorporated into rock.

The primary biological importance of phosphates is as a component of nucleotides, which serve as energy storage within cells (ATP) or when linked together, form the nucleic acids DNA and RNA. Phosphorus is also found in bones, whose strength is derived from calcium phosphate, and in phospholipids (found in all biological membranes).

Phosphates move quickly through plants and animals; however, the processes that move them through the soil or ocean are very slow, making the phosphorus cycle overall one of the slowest biogeochemical cycles.

Human Influence

Artificial fertilizers and other wastes not absorbed by plants mostly enter the groundwater and collect in streams, lakes and ponds. The extra phosphates are a major contributor to the process called eutrophication, which causes excessive growth of water plants and algae populations and subsequent depletion of dissolved oxygen potentially suffocating fish and other aquatic fauna.

Sulphur Cycle

Sulphur is one of the constituents of many proteins, vitamins and hormones. It recycles as in other biogeochemical cycles.

The essential steps of the sulphur cycle are:

- Mineralisation of organic sulphur to the inorganic form, hydrogen sulphide: (H_2S).
- Oxidation of sulphide and elemental sulphur (S) and related compounds to sulphate (SO_4^{2-}).
- Reduction of sulphate to sulphide.
- Microbial immobilisation of the sulphur compounds and subsequent incorporation into the organic form of sulphur.

Bibliography

Alfoldi, L.: *Hydrogeologic Considerations for in situ Bioremediation,* Butterworth-Heinemann Publishers, Stoneham, 1991.

Anderson, T. A. and Coats, J. R.: *Bioremediation Through Rhizosphere Technology,* American Chemical Society, Washington, 1994.

Antebi, Elizabeth, and David Fishlock: *Biotechnology, Strategies for Life,* Massachusetts, Cambridge, 1986.

Bailey, Ronald: *Ecoscam, the False Prophets of Ecological Apocalypse,* St. Martin's Press, New York, 1993.

Barley, S., J. Freeman and R. Hybels: *Strategic Alliances in Commercial Biotechnology,* Harvard Business School Press, Boston, 1992.

Batra, R. and W. Klassen: *Public Perception of Biotechnology,* Agricultural Research Institute, Maryland, 1987.

Biotechnology Industry Organization (BIO): *Agriculture and the New Biology,* The Association, Washington, 1987.

Biotechnology Research Subcommittee: *Committee in Fundamental Science and Research,* National Science and Technology Council, Washington, 1995.

Blakely, E.: *The Economic Development Potentials of California's Biotech Industry, Biotech Industry Research Group, Institute of Urban and Regional Development,* University of California, Berkeley, 1989.

Blakely, Edward J. and Kelvin W. Willoughby: *Institute of Urban and Regional Development, Biotechnology Industry Research Group,* University of California, Berkeley, 1990.

Blum, Deborah: *The Monkey Wars,* Oxford University Press, London, 1994.

Booz, Allen & Hamilton: *Foreign Investment in Bay Area Bioscience,* Bay Area Bioscience Center, Oakland, 1991.

Center for Agricultural Molecular Biology: *Cook College, New Jersey Agricultural Experiment Station, Rutgers,* The State University, New Brunswick, 1991.

Chong, Ken P. and Smith, John W.: *Mechanics of Oil Shale, London, England and New York,* Elsevier Applied Science Publishers, New York, 1984.

Cirpka, O. and Bisch, G.: *Technical Design of the Aquifer Model and Influence of Spatial Variability,* Presented at the Third International Symposium on In-Situ and On-Site Bioreclamation, California, 1995.

Committee on In Situ Bioremediation Water Science and Technology Board Commission on Engineering and Technical Systems National Research Council: *In Situ Bioremediation, When does it Work,* National Academy Press, Washington, 1993.

Committee on Scientific Evaluation of the Introduction of Genetically Modified Microorganisms and Plants into the Environment: *Field Testing Genetically Modified Organisms, Framework for Decisions,* National Academy Press, Washington, 1989.

Comstock, G.: *What Obligations have Scientists to Transgenic Animals, College Station, Center for Biotechnology Policy and Ethics,* Texas A&M University, *New Jersey,* 1992.

Cooney, J. J. and Gadd, G. M.: *Metals and Microorganisms, Journal of Industrial Microbiology,* Washington, 1995.

Crawford, Ronald L. and Don L.: *Crawford, Bioremediation*

Principles and Applications, Cambridge University Press, New York, 1996.

Dale, Jeremy W.: *Molecular Genetics of Bacteria*, John Wiley and Sons, New York, 1989.

De Lorenzo, V.: *Genetic Engineering Strategies for Environmental Applications*, Current Opinion in Biotechnology, California, 1992.

Drexler, K. Nanosystems: *Molecular Machinery, Manufacturing, and Computation*, John Wiley and Sons, New York, 1992.

Economidis, I. and Palmisano A.: *Environmental Biotechnology, EC-US Task Force on Biotechnology Research*, Brussels, Belgium, 1995.

Ehrlich, H. L.: *Geomicrobiology*, Marcel Dekker, New York, 1990.

Esser, K.: *Plasmids of Eukaryotes, Fundamentals and Applications*, Springer-Verlag Berlin Heidelberg, Germany, 1986.

Fowle, John R.: *Application of Biotechnology, Environmental and Policy Issues*, Westview Press, Colorado, 1987.

Francis, A. J.: *Microbial Dissolution and Stabilization of Toxic Metals and Radionuclides in Mixed Wastes*, Oxford, New York, 1990.

Frankenberger, W. T. and Benson S. M.: *Selenium in the Environment*, Marcel Dekker, New York, 1994.

Friedman, D.: *The New Economy Project Final Report*, The New Economy Team, UCLA, 1994.

Gendel, Steven: *Agricultural Bioethics, Implications of Agricultural Biotechnology*, Iowa State University Press, Ames, 1990.

Gibson, D. T. and Sayler, G. S.: *Scientific Foundations of Bioremediation, Current Status and Future Needs*, American Academy of Microbiology, Washington, 1992.

Ginzburg, Lev. R.: *Assessing Ecological Risks of Biotechnology*, Butterworth-Heinemann, Boston, 1991.

Goodsell, David S.:*The Machinery of Life,* Springer-Verlag, New York, 1993.

Grimes, D. J.: *Validation of Reliability and Predictability of Environmental Biotechnologies,* Workshop on Bioremediation, Japan, 1995.

Grocery Manufacturers of America: *Safe Foods Developed Through Biotechnology, Nutritious, Healthful,* Abundant and Good Tasting, 1993.

Hacking, Andrew J.: *Economic Aspects of Biotechnology,* Cambridge University Press, Cambridge, 1986.

Harkness, M. R., McDermott J. B. and Abramowicz D. A.: *In Situ Stimulation of Aerobic Degradation of PCB Biodegradation in Hudson River Sediments,* MacMillan, London, 1993.

Hettinger, N.: *Owning Varieties of Life, Biotechnology, Intellectual Property, and Environmental Ethics, Center for Biotechnology Policy and Ethics,* Texas A&M University, Texas, 1994.

Hettinger, N.: *Biotechnology Policy and Ethics,* A&M University, Texas, 1992.

Hinchee, R. E., Anderson D. B., Metting, F. B. and Sayles G. D.: *Applied Biotechnology for Site Remediation,* Lewis Publishers, Florida, 1994.

Hinchee, Robert E.: *Applied Biotechnology for Site Remediation,* CRC Press, Ann Arbor, 1994.

Hollenberg, C. P. and H. Sahm: *Microbial Genetic Engineering and Enzyme Technology,* Gustav Fischer, New York, 1987.

Kelly, Kevin: *Out of Control,* Addison-Wesley, Menlo Park, 1994.

Kenney, Martin *Biotechnology: The University-Industrial Complex,* Yale University Press, Connecticut, 1986.

Kimbrell, Andrew: *The Human Body Shop,* Harper/Collins, San Francisco, 1993.

King, R. B., Mihony, G. and Sheldon J. K.: *Practical Environmental Bioremediation*, Lewis Publishers, Florida, 1992.

Kloppenburg, Jack Ralph: *The Political Economy of Plant Biotechnology*, Cambridge University Press, Cambridge, 1988.

Koehler, Gus.: *Tools to Promote Economic Success*, California Research Bureau, Sacramento, 1995.

Kotkin, Joel and Levy, Stephen: *A Twenty-First Century Prospectus, Center for the New West, and Center for the Continuing Study of the California Economy*, California, 1996.

Krimsky, Sheldon: *Biotechnics and Society,The Rise of Industrial Genetics*, Praeger, New York, 1991.

Lacy, W., L. Busch, and L. Lacy: *Agricultural Biotechnology, Issues and Choices*, Purdue University Press, Indiana, 1991.

Lacy, W., L. Lacy and L. Busch: *Emerging Trends, Consequences, and Policy Issues in Agricultural Biotechnology*, Westview Press, Colorado, 1991.

Lee, V. and Yen T.F.: *Biopolymer Shields, in Applied Biotechnology for Site Remediation*, Lewis Publishers, Florida, 1994.

Levorsen, A. I.: *Geology of Petroleum, San Francisco*, W. H. Freeman and Company, California, 1967.

Lovley, D. R.: *Dissimilatory Metal Reduction*, Annual Review of Microbiology, 1993.

Lynch, J. M. and Hobbie J. E.: *Concepts and Applications in Microbial Ecology*, Blackwell Scientific Publications, England, 1988.

McKinsey Global Institute: *Manufacturing Productivity*, McKinsey and Company, Washington, 1993.

Metting, Jr., F. B.: *Soil Microbial Ecology, Applications in Agricultural and Environmental Management*, Marcel Dekker, New York, 1992.

Miller, Henry I., M.D : *Remarks for the California Legislature*

Assembly Committee on Agriculture, MacMillan, London, 1994.

Miller, R. V. and Poindexter J. S.: *Strategies and Mechanisms for Field Research in Environmental Bioremediation,* American Academy of Microbiology, Washington, 1994.

Mueller, J. G.: *Model Studies of Bioremediation Strategies and Monitoring Technologies in Proceedings, Environmental Biotechnology,* EC-US Task Force on Biotechnology Research, Belgium, 1995.

National Research Council (NRC): *Alternatives for Ground Water Cleanup,* National Academy Press, Washington, 1994.

Old, R. W. and S. B. Primrose: *Principles of Gene Manipulation, An Introduction to Genetic Engineering,* University of California Press, California, 1980.

Olson, Steve: *An Industry Comes of Age,* National Academy Press, Washington, 1986.

Organisation for Economic Cooperation and Development (OECD): *Biotechnology for a Clean Environment, Prevention, Detection, Remediation,* OECD, Paris, 1994.

Ouellette, Robert P. and Paul N.: *Cheremisinoff, Essentials of Biotechnology, Lancaster,* Technomic Publishing Company, Pennsylvania, 1985.

Panem, Sandra: *Biotechnology, Implications for Public Policy,* The Brookings Institution, Washington, 1985.

Powell, W. and P. Brantley: *Competitive Cooperation in Biotechnology, Learning Through Networks,* Harvard Business School Press, Boston, 1992.

Prentis, Steve: *Biotechnology, A New Industrial Revolution,* George Braziller, New York, 1984.

President's Council on Competitiveness: *Report on National Biotechnology Policy,* The Council, Washington, 1991.

Rissler, Jane, and Margaret Mellon: *Perils Amidst the Promise,*

Ecological Risks of Transgenic Crops in a Global Market, Union of Concerned Scientists, Cambridge, 1993.

Saaty, R. P. and Booth S. R.: *Cost Effectiveness of a Remediation Technology Field Tested at the Savannah River Integrated Demonstration Site*, Los Alamos National Laboratory, New Mexico, 1994.

Saunders, Venetia A. and Jon R. Saunders: *Microbial Genetics Applied to Biotechnology, Principles & Techniques of Gene Transfer & Manipulation*, Crook Helm Ltd., London, 1987.

Saxenian, A.: *Regional Advantage*, Harvard University Press, Boston, 1994.

Sayler, G. S., Fox, R. and Blackburn, J. W.: *Environmental Biotechnology for Waste Treatment*, Plenum Press, New York, 1991.

Sayler, G. S.: *Biodegradation Process Analysis, Molecular Application in Simulations and Environmental Verification in Proceedings, Environmental Biotechnology*, EC-US Task Force on Biotechnology Research, Belgium, 1995.

Sayler, Gary S.: *Environmental Biotechnology for Waste Treatment*, Plenum Press, New York, 1991.

Skladany, G. J. and Metting, F. B.: *Bioremediation of Contaminated Soil in Soil Microbial Ecology*, Marcel Dekker, New York, 1992.

Smith, George P.: *The New Biology, Law, Ethics and Biotechnology*, Plenum Press, New York, 1989.

Sussman, M.: *The Release of Genetically-engineered Micro-organisms*, Harcourt Brace Jovanovich, New York, 1988.

Thomas, J. M. and Ward, C. H.: *In Situ Biorestoration of Organic Contaminants in the Subsurface*, Environmental Science and Technology, 1989.

Thompson, P: *Animal Welfare and Animal Rights, Center for Biotechnology Polilcy and Ethics*, Texas A&M University, Texas, 1992.

Thompson, Paul B., Robert J. Matthews: *Ethics, Public Policy and Agriculture*, MacMillan, London, 1994.

Tunlid, A. and White, D. C.: *Biochemical Analysis of Biomass, Community Structure, Nutritional Status, and Metabolic Activity of Microbial Communities in Soil in Soil Biochemistry*, Marcel Dekker, New York, 1992.

Wijesinghe, A. M., Knapp, R. B. and Carman, L. M.: *Preliminary Feasibility and Cost Analysis of the In-Situ Microbial Filter Concept*, Lawrence Livermore National Laboratory, Livermore, California, 1992.

Index

C

K

L

M

N

O

T

V

W

X

Z

❑❑❑